The Political Economy
of Agricultural Booms

Mariano Turzi

The Political Economy of Agricultural Booms

Managing Soybean Production in
Argentina, Brazil, and Paraguay

Mariano Turzi
Universidad Torcuato Di Tella
Buenos Aires, Argentina

ISBN 978-3-319-45945-5   ISBN 978-3-319-45946-2 (eBook)
DOI 10.1007/978-3-319-45946-2

Library of Congress Control Number: 2016951447

Cover illustration: Pattern adapted from an Indian cotton print produced in the 19th century

Printed on acid-free paper

This Palgrave Macmillan imprint is published by Springer Nature
The registered company is Springer International Publishing AG
The registered company address is: Gewerbestrasse 11, 6330 Cham, Switzerland

To my wife Gladys, my one true love. Ab imo pectore

PREFACE

This book studies the *international political economy* of *agriculture*, specifically of the global agroindustrial model of soybean production and the domestic variations across three *national* case studies: Brazil, Argentina, and Paraguay (the BAP countries).

Chapter 1 introduces *agriculture* and presents the analytical framework. It begins with an empirical, historical background on soybeans and the world soy market. It then proceeds to study agriculture and its linkages to the economy, reviewing the main debates and recent contributions in the political economy of agriculture literature. The final section studies the several trends in global demand that have come together to intensify competition for agricultural resources and food products. World demand for agricultural commodities is driven by four factors (the "four f's"): food, feed, fuel, and finance.

The international political economy structure of agriculture is currently a corporate-driven, vertically integrated system of global agricultural production. This is the result of two mutually reinforcing traits: the technological transformation into agrochemicals and genetically modified seeds and the economic globalization of grain trading. The two sections review the supply-side actors who have driven this international restructuring of production and trade: chemical and trading multinational corporations (MNCs). Chapters 2 and 3 focus on the *international* dimension from the supply side. The growing importance of information technologies and biotechnology has led to a dramatic increase in the power of the seeding companies within the soybean chain. The strategic value of a unique asset—genetically modified seeds with proprietary traits—has propelled these companies to a dominant position. The power of input suppliers in the new soybean mode of production has given them overriding influence, allowing them to appropriate a sizeable portion of the rents generated along the chain. The importance of tracing the behavior of these multinational corporate actors lays in the fact that they have exerted their power to create the institutional structure to govern the new resource (genetically modified soybeans). As such, the resulting institutional landscape is a "map" that exhibits the marks of the power struggles between the actors in the chain in their attempt to crystallize their power resources into the governing structure, objectifying their power (O'Donnell 1978). A mirror situation can be found in Chapter 3 at the level of corporate actors in the trading and industrial processing stage. Distributors and processors have taken advantage of the grain trade liberalization of the last decade to leverage their position in open markets. They concentrated on supply mechanisms through the advantages derived from scale and vertical integration. Their strategies for furthering their position within the soybean chain include infrastructure development, financial leveraging, and flexible sourcing.

Transnationalization is increasingly eroding the relevance of national frontiers. The mode of production in the soybean chain would be thus regionalized according to corporate incentives operating in a global trend toward relocation of the different stages of production. Soybeans are harvested in Paraguay, sent by barge to Brazil or Argentina for processing, and sold in Geneva to Asia after headquarter in the USA has authorized the operation. The geoeconomic pull of the international-level corporate strategies is reorganizing territorial boundaries, integrating the three countries into a single regional production structure from the upstream to the downstream: the "Soybean Republic" (2011). The international model of agricultural production has empowered chemical and trading multinational companies. The vertical integration of these two powerful links of the globalized chain has generated a commanding production structure. To consolidate this "soybean complex" of production, chemical companies have used their scientific and technological superiority to advance the sales of their agrochemical products. They have integrated with traders and processors and leveraged scale advantages to establish dominant buying positions. Further, they have drawn on their financial strengths to dictate infrastructural developments, thus creating a pull force to rearrange the economic geography through the BAP countries. Indeed, the analysis of the trading link in the soybean chain evidences that national borders were becoming increasingly irrelevant realities. The ascent of an international, corporate-driven model of organization of production is reshaping territorial realities according to global production demands and needs.

But—as Harris (2001) points out—modes of production evolve from the contradiction between means (material forces) of production and the (social) relations of production. A mode of production encompasses the totality of the social and technical human interconnections involved in the social production and reproduction of material life. The material underpinnings of social cleavages (Lipset and Rokkan, 1967) in the agricultural sector are different in each country. These cleavages have impacted policy response, generating specific national political economy configurations. The reality that international-level stimuli impacted domestic institutional structures in Brazil, Argentina, and Paraguay is the explanatory core of this book. The means of production changed for the three countries in a similar way, but the relations of production did not because conditions on the ground differ significantly in Argentina, Brazil, and Paraguay. Despite uniformity of the international corporate actors driving the process of soybean expansion, the end results were far from homogeneous. They reinforced the existing structures of power (concentrated landowners) in Paraguay but upset the balance of power (between the urban/industrial/labor coalition and the agricultural interests) in Argentina, while they empowered local actors (municipalities and state governments) in Brazil. Results are not preordained by economic factors—as modernization theory would predict—nor is there evidence of convergence of processes due to globalization. Each of the BAP exhibits different patterns of institutional governance of the soybean chain, and the level of centralization of resource management gives the basis of comparison among the three case studies. The application of a comparative political economy analysis reveals rather the ascent of "Soybean Republic." National coalitions have limited the convergence and standardization associated with economic globalization (Guillén, 2009). The end result observed in the case studies of Brazil, Argentina, and Paraguay—the BAP countries—demonstrates a key role of national political economy arrangements in shaping the influence of the globalizing "pull" forces. Pressure groups and coalitions have been formed around agricultural interests, and their relative strength has been the determinant factor transforming natural endowments in these three countries into competitive advantages in world markets. The preferences and relative power of actors within these societies—economic and political, national and subnational, public and private—within certain institutional and policymaking frameworks are giving way to differential patterns. Chapters 4, 5, and 6 analyze the *political economy* of producers and farmers throughout the three case studies. If the global structure conditions explain the *why*, the comparative political-economic case-study analysis of domestic political economy structures in Brazil, Argentina, and Paraguay accounts for the *how*. The diverse cleavages and institutional forms throughout the BAP have resulted in specific, non-convergent modes of production for the same natural resource. In each of the case studies, the changes in the means of production have created different— although not exclusive—relations of production. The focus is not on the agronomic component of soybean production, but rather on the broader set of sociopolitical and socioeconomic issues surrounding it. This book is less concerned with the increasing physical space or economic weight of a crop and more with the expansion and consolidation of control structures and social relations. The analysis of soybean production is treated as a heuristic device to expose the underlying balance of power of the actors in the chain and the way in which they have adapted to and shaped the institutional structure governing resource production and allocation. Different institutional settings and governance rules will give rise to different forms of resource administration. This is the guiding question in this book: what have been the effects of different governing institutions (in Argentina, Brazil, and Paraguay) on the management of a resource and export product (soybeans). The Brazilian case is one in which local governance is much stronger, which has allowed to effectively integrate state institutions with the resource/sector (*coordination*). In Paraguay, although the formal structure is that of a unitary state, the agricultural sector has achieved de facto decentralization by state capture. Taking advantage of power asymmetries and weak initial institutional conditions, there has been *colonization* by particular and foreign interests. Finally, Argentina is a case of centralized institutions exhibiting a conflictive pattern of relations with the economic sector/resource (*confrontation*).

The concluding chapter reviews research findings and poses demanding questions for future international political economy research, pressing public policy dilemmas for nation-states.

ACKNOWLEDGMENTS

I would like to thank Director of the Latin American Studies Program at the Johns Hopkins University's Paul H. Nitze School of Advanced International Studies Riordan Roett for trusting in my subject study and me. He is the "intellectual landowner" of the Soybean Republics. Norma González and her support through the Fulbright scholarship were of key importance as well.

For the past fifteen years, I have had the honor and privilege of a true mentor like Sergio Berensztein. Among the countless personal and professional debts of gratitude I owe to him, "sowing the seeds" of this area of study is the one most directly related to this book.

I have also had the guidance and permanent support of Roberto Russell. I am very thankful to Torcuato Di Tella University Rector Ernesto Schargrodsky and PoliSci/IR Department Directors Catalina Smulovitz and Juan Tokatlian for giving their vote of confidence. A recognition is also in order for former Business School Director Juan José Cruces and MBA Director Sebastián Auguste.

To Palgrave editor Dr. Anca Pusca, who trusted in this project, and to Juan Pablo Luna. in representation of REPAL (Network for the Study of Political Economy in Latin America). They have delivered on the promise of promoting new studies in the political economy of Latin America and welcoming innovations that challenge the conventional wisdom on socially relevant phenomenon in the region with an open and eclectic approach.

I would like to acknowledge the high-quality data and material from the United States Department of Agriculture's Foreign Agricultural Service (FAS-USDA) and the generosity of the Production Estimates and Crop Assessment Division (PECAD) to share them.

Finally, I am grateful to Lester Brown, Harry De Gorter, Gary Gereffi, Jeffrey Sachs, Carlos Scartascini, Ernesto Stein, Johan Swinnen, Mariano Tommassi, Steven Topik, and Tom Vilsack for their invaluable insights. Also, Lucio Castro, Blairo Maggi, Gustavo Grobocopatel, and Fernando Lugo for their interest, time, and comments.

CONTENTS

1 The International Political Economy of Agriculture   1
   Driving Demand: The Four F's   1
   Soybeans and the World Market   4
   Agriculture in the Latin American Economies   8
   Soybeans in the Southern Cone   11
   *Linkages, Commodity Chains, and the Political
   Economy of Agriculture*   14

2 A Super-Seeding Business   23
   The Institutional Frameworks   28
   The Political Economy of Seeds   31
     Argentina   32
     Paraguay   36
     Brazil   40

3 Global Trading   49
   The New Global Agricultural Trade   50
   World Grain Trade and the Soybean Chain   55
   Finance and Infrastructure as Political Economy   66
     Financial Instruments   67
     Taxes/Duties   70
     Infrastructure   74

4 Coordination: Brazil   83
The Amazon: Political Economy in Brazil's Far West   85
Land Struggles   91

5 Colonization: Paraguay   101
The Brasiguayos: An Intermestic Driving Force   105
The Far West   109

6 Confrontation (. . . and Beyond): Argentina   117
A State Against the Campo?   118

Conclusion   127

**Annex 1. Agricultural Forward
and Backward Linkages**   131

**Annex 2. Geographic Distribution
of Soybean Production in the Soybean Republics**   137

References   141

Index   149

LIST OF FIGURES

Fig. 1.1   Soybean meal and farming industry, world, 1976–2015   7
Fig. 2.1   The seed production circuit   31
Map A.1   Brazil: soybean production by state   137
Map A.2   Argentina: soybean production by province   138
Map A.3   Paraguay: soybean production by province   139

LIST OF TABLES

Table 1.1   Evolution of soybeans in Argentina, Brazil,
             and Paraguay from 1985/1986 to 2015/2016   13
Table 1.2   World share, soybean exports of Argentina, Brazil,
             and Paraguay from 1985/1986 to 2015/2016   14
Table 3.1   Tax rates for soybeans, *Retenciones móviles* (2008)   72

CHAPTER 1

The International Political Economy of Agriculture

Abstract Introduction of *agriculture* presents the analytical framework. It begins with an empirical, historical background on soybeans and the world soy market. It then proceeds to examine the literature on agriculture and its linkages to the economy, reviewing the main debates and recent contributions in the political economy of agriculture literature. The final section studies the several trends in global demand that have come together to intensify competition for agricultural resources and food products. World demand for agricultural commodities is driven by four factors (the "four f's"): food, feed, fuel, and finance.

Keywords Agriculture · Soybeans · Agribusiness · International political economy · Latin America · Commodity chains · Commodities · Development

Driving Demand: The Four F's

Several trends in global demand have come together to intensify competition for agricultural resources and food products. World demand for agricultural commodities is driven by four factors (the "four f's"): food, feed, fuel, and finance.

The first factor, food, results from a demographic dynamic: the global population grows by around 80 million people per year. The first billion was reached in 1804. Owing mainly to technological advances in the fields

© The Author(s) 2017
M. Turzi, *The Political Economy of Agricultural Booms*,
DOI 10.1007/978-3-319-45946-2_1

1

of medicine and agriculture, from that point until 2014 the world population grew more than 600 percent, to more than 7 billion. In 2009 the renowned agronomist Norman Borlaug estimated that over the next fifty years, the world would have to produce more food than it had in the past 10,000 years.[1] The World Bank projected in April 2016 that food demand would rise by 20 percent globally over the next fifteen years. The compounded result: more people in the world, living longer, means a structural upward shift in food demand. Moreover, the world population is changing not just quantitatively but also qualitatively. India and China have the largest rural populations, 857 million and 635 million, respectively. However, they are also expected to experience the largest declines in rural residents, with a 300 million reduction in China and a 52 million reduction in India anticipated by 2050. In 2010, for the first time, more than half of the world's population was urban. By 2014, the total urban population had grown to 54 percent, and this share is expected to increase to 66 percent by 2050. The UN's Population Division 2014 projections indicate that India is expected to add more than 11 million urban dwellers every year and China more than 8 million.[2]

The second driver of agricultural demand, feed, is mostly attributable to the rise of the emerging world, with a regional focus on Asia, particularly on China and India. Global poverty rates started to fall by the end of the twentieth century largely because emerging countries' growth accelerated from average annual rates of 4 percent in 1960–2000 to 6 percent in 2000–2010. Around two-thirds of poverty reduction within a country comes from growth, and greater equality contributes the other third. According to a World Bank estimate, between 2005 and 2012, India lifted 137 million people out of poverty.[3] For China, the World Bank calculates that, from the time market reforms were initiated in 1978 until 2004, the figure rose to more than 600 million, and in more recent years (between 2005 and 2011), nearly 220 million people have been lifted out of poverty.[4] When living standards rise, so does the demand for meat and dairy products. As people from China and India abandon poverty and move into the burgeoning global middle class—in Asia alone, the figures for 2014 were estimated at 500 million, and they are projected to surpass 3 billion by 2030—they diversify their diets to include more vegetable oils, meat, and dairy products. Not only are there more people to feed, but more people are eating pork, chicken, and beef.

Against this backdrop, soybeans become the most essential input in the global food system. The bean contains 83 percent flour and 17 percent oil.

When oil is extracted, the remaining residue is known as soybean cake, meal, or pellets; it is a vegetable protein concentrate (42–44 percent). Meal has found its strongest application as fodder for the industrial raising of farm animals, or "factory farming." Soybeans can also be processed for human consumption in a variety of forms: as soy meal, soy flour, soy milk, soy sauce, tofu, textured vegetable protein (found in a variety of vegetarian foods and intended to substitute for meat), lecithin, and oil. Soybean oil is the world's most widely used edible oil and has several industrial applications. Soybeans are thus a highly efficient crop: about 40 percent of the calories in soybeans are derived from protein, compared to 25 percent for most other crops. This means that the return per dollar spent is relatively high compared to that for other oilseeds.

In the lower-income segments, soy is an essential component of any dietary energy supply intended to inexpensively cover daily calorie requirements. For the better off, the crop is a cornerstone fodder component. As livestock can be fed more efficiently with soybean-based feed, the massive spread of the crop has made chicken, beef, and pork cheaper and more readily available worldwide. According to estimates from the US Department of Agriculture (USDA), China and India are the world's top importers of soybean oil and are projected to remain so in the coming years.[5] China tops current importing charts and projected scenarios as soybean importer; its soybean imports were projected to reach 72 million tons (MT) in 2014–2015, meaning that China alone was expected to absorb 64 percent of total global soybean exports by that year.

The third factor pushing up demand for grain production is fuel. The first explanation is that the price of oil has a direct impact on prices of agricultural inputs such as fertilizers. When the price of fossil fuels rises, then it becomes a rational economic alternative to divert food crops into the production of biofuels. The debate about peak oil and the subsequent expectations of oil price hikes—plus the risk of supply shortages—have triggered a growing demand for energy from the biofuels industry. Supported by policy mandates, countries are seeking to diversify their energy sources by incorporating renewables. The Food and Agriculture Organization (FAO) estimated in 2013 that biofuel prices would continue to rise—16–32 percent higher in real terms compared to the previous decade—over the next ten years, with expected high crude oil prices and continuing biofuel policies around the world that promote demand.

The financial component of agricultural demand is more indirect and more controversial, but nevertheless, it is equally important in light of the speculation in food commodity markets, particularly by institutional investors such as hedge funds, pension funds, and investment banks. Since 2000 there has been a fifty fold increase in dollars invested in commodity index funds. The number of commodity futures contracts outstanding nearly doubled between 2004 and 2007. However, commodity prices crashed with equities following the financial crisis and traded tightly in line with the stock market over the nervous years that followed, providing no diversification. After 2005 commodities did begin to move more closely in line with other asset classes and with each other. This became especially close during the financial crisis. After the 2008 financial crisis, global investors seeking safe hedges for their portfolios in the face of depreciation of the US dollar turned commodities into an asset class. The correlation between commodities and stocks—negative before—became strongly positive. But 2010 was the last year investors pumped net cash into commodity index swaps. Outflows trickled, becoming an outpour in 2014, when the value of commodity assets under management was reduced $24.2bn to a total of $67bn from a pre-crisis high of more than $150bn. The financialization of commodity markets is self-perpetuating: as new investment products—food derivatives and indexed commodities—create speculative opportunities in grains, edible oils, and livestock, prices for food commodities increase. More money flows into the sector, and a new round of price increase follows. Although food inflation and food volatility have increased alongside commodity speculation, there is no conclusive evidence of the impact of finance as a driver of price developments. The UN Conference on Trade and Development 2009 Report stated that index traders "can significantly influence prices and create speculative bubbles, with extremely detrimental effects on normal trading activities and market efficiency," something supported by the research done by Tang and Xiong (2010), who found that financialization made ostensibly different commodities such as grains and oil more closely correlated after 2004, relating the trend to "large inflows of investment capital to commodity index securities during this period." However, Bhardwaj et al. (2015) argue that the impact of financialization was marginal.

SOYBEANS AND THE WORLD MARKET

Soybeans (US) or soyabeans (UK) are the common denomination of the *Glycine max*. The English word *soy* derives from the Chinese *shu* and the Japanese *shōyu* (soy sauce), and *soya* comes from the word's Dutch adaptation. This legume was first cultivated in northern China and spread into Japan, Korea, and the rest of Southeast Asia during the Chou Dynasty. Known to the Chinese for 5,000 years, soybeans were one of the five "sacred seeds," together with barley, millet, rice, and wheat. According to Chinese tradition, the first written record of the crop dates from 2838 B.C., when Chinese emperor Sheng-Nung—The Heavenly Farmer—writes in his *Materia Medica* about soy's medicinal properties.[6] Although soybeans remain a crucial crop in China, Japan, and Korea, today only 45 % of world production is located in Asia. The other 55 % percent of production is in the Americas, divided mainly between the USA, Brazil, and Argentina. Soy was first researched in Europe in 1712 by Englebert Kaempfer, a German botanist who had studied in Japan. The first seeds were planted in the *Jardin des Plantes*, Paris in 1740. Swedish botanist Carl von Linne made the first scientific study of the soybean in the West, giving it its scientific name due to its large nitrogen-producing nodules on its roots. In the early nineteenth century, trading ships first introduced soybeans in the Western Hemisphere, where it was considered an industrial product. Even Henry Ford promoted the soybean, producing auto body panels made of soy-based plastics.[7]

The plant is usually between 40 and 140 cm tall. The fruit is a hairy pod of 3–8 cm that contains three to five beans. Cultivation is successful in climates with hot summers, with optimum growing conditions in mean temperatures of 20°–30°C (68°–86°F). The crop grows in a wide range of soils, with optimum growth in moist alluvial soils. In symbiosis with the bacterium *Bradyrhizobium japonicum*, the plant fixes nitrogen to the soil, allowing for a beneficial biological cycle that slows down the soil degradation. Nitrogen is found mainly in the stubble, which remains in the ground after the harvest, making it as the crop's own "green" fertilizer.

Classified as an oilseed, soy is cultivated for its beans and to extract oil. The bean is an important source of protein (35 %), which is why it has long been considered the basis of the food pyramid for peoples with scarce access to proteins from animal sources. The bean contains 83 % flour and 17 % oil. When oil is extracted, the remaining residue is known as soybean cake, meal, or pellets—a vegetable protein concentrate (42–44 %). Meal has found its strongest application as fodder for the industrial raising of farm animals or "factory farming." Soybeans can also be processed for human consumption in a variety of ways: soy meal, soy flour, soy milk, soy sauce, tofu, textured vegetable protein (found in a variety of vegetarian foods intended to substitute for meat), lecithin, and oil.[8] Soybean oil is the world's most widely used edible oil and has several industrial applications. By mid-twentieth century, a combination of factors that included demographics, technology, economics, and international conflagration began to alter the shape of rural production. Prior to World War II, most livestock and poultry came from family farms. Cattle were usually grazed on rangeland or pasture and were fed hay, silage, and some corn during the winter. Poultry flocks were small and ate barnyard scraps. Since open range grazing[9] was only possible in the great land extensions of the New World, livestock farming experienced a drastic transformation. Cattle began to be kept in large, insulated structures (stall barns and loafing barns) and were fed a mix of root crops and grain. Although farmers had been using mixed feeds—grains, oilseed meals, etc.—in small quantities since the late 1800s, their use accelerated in the late 1930s with scientific feed formulation and the discovery of essential amino acids, protein complementarity, and the concept of animal nutrition. Scientific feed formulation designed to maximize animal growth at the least cost favored the use of soybean meal as a protein source. During the 1940s and 1950s, the centralized, low-cost feedlot infrastructure combined with (soybean-based) fortified and balanced feeds produced more efficient and profitable livestock and poultry. Feedlots also helped to allocate the feed grains surplus from the 1950s by converting it into profitable meat products (Shurtleff and Aoyagi 2007; Part 7). The chemical industry developed fertilizers that replaced animal manures, so animals were no longer needed on the farm. Labor-saving mechanization encouraged production centralization and automation, converting the farms into "animal factories." Soybeans became a key input for this feedlot mode of production. Not only did soy have high protein content, but soybean meal and surplus feed grains were also initially very low in cost. In fact, the evolution of soybeans is intimately related to the rise in animal protein consumption worldwide, which only became possible with confined farming techniques, of which soybean is the cornerstone. Soybeans are a highly efficient crop: the total cost of the crop is relatively low compared to its unit proteic value. About 35–38 percent of the calories in soybeans are derived from protein, compared to 20–30 percent in most other beans. Indeed, according to the American Soybean Association *Soystats* 2015 Report, soybeans represent 68 % of world protein meal consumption, followed very distantly by rapeseed (14 %) and sunflower (6 %). This means the "proteic return" per dollar spent is relatively higher compared to other oilseeds or fodder components. As a result, soybeans played an increasingly important role as a food source for an even larger segment of a changing farm animal population. Poultry is more efficient than swine or beef in converting feed to meat, in terms of cost and time. On average—depending on the composition of the feed, which technological advances modify almost monthly—it takes about 3 kg of feed protein to produce 0.45 kg of broiler protein. To produce the same amount of pork protein requires 3.77 kg and for the equivalent beef protein 6.5 kg of feed are required.[10] Integration and automation led to scale returns, and overall efficiency gains lowered poultry prices by mid-1970s. This sustained rise in consumption has been a major source behind the steady rise of soybean production, as the following figure shows Fig. 1.1.

On the *supply side*, the initial takeoff of soybean demand coincided with the collapse of a major substitute—the Peruvian anchovy—in the early 1970s, due to El Niño and over-fishing.[11] This depletion led to a major decline of high-protein feedstock and to a decision to switch to the more cost-efficient soymeal as a protein source. The European Community (EC), a major soybean importer since World War II, lifted trade restrictions. In the 1960 Dillon Round of the GATT, the EC had agreed to a zero tariff binding on soybeans and to low tariffs on soy-derived products, increasing the international demand for soybeans and soy cake. These new market conditions in Europe also acted an incentive to production in South America. At the same time, the USA, Australia, Canada, and the USSR experienced production shortfalls due to adverse weather conditions, which persisted for several years. Although the boom period would

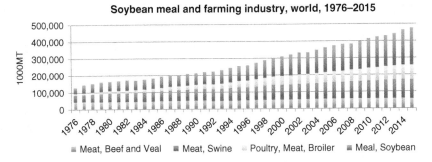

Fig. 1.1   Soybean meal and farming industry, world, 1976–2015

Source: Author's calculation based on USDA data. MT = Metric Tons

not be for another fifteen years, the combination of these factors stimulated oilseed production in the Southern Cone.

On the *demand side*, during the 1970s the Soviet Union and other centrally planned economies entered into the global grain markets, with a significant effect on the grain and oilseed trade. Abundance of oil revenues (petrodollars) meant availability of credit to help finance global trade growth. By 1980s, China was opening up to world trade, and the export-led Asian model of development, epitomized in the four tigers,[12] was being showcased as a model of success. With the improvement in living standards throughout Asia, the demand for meat and dairy products grew as well. The demand for agricultural food commodities has been steadily growing in emerging economies, as bourgeoning middle and upper classes diversify their diets to include more vegetable oils, meat, and dairy products. As a result, developing countries' demand for grains and oilseeds for livestock feed has risen disproportionately, rising more quickly than overall demand for food. According to USDA data, domestic worldwide consumption for soybean oil increased 531 % for the period 1975/1976 to 2015/2016. For the same period and product, the percent increase in Southeast Asia rose to 1511 %, in East Asia 1866 %, and in South Asia 2908 %. While for the same period the world domestic consumption of soybean meal increased 441 %, in China alone the increase was 5676 %.

Agriculture in the Latin American Economies

The role of the agricultural sector has been to some extent overlooked in the macroeconomy of the BAP countries for the last fifty years. Instead of capitalizing on a relatively abundant natural resource endowment and its resultant competitive advantage, policymakers have used the sector as a cash cow to be milked in order to subsidize relatively more inefficient—yet politically more attractive—domestic industrial sectors. This "bias against agriculture" has explanations at many different levels. After the first third of the twentieth century, a consensus began to emerge among economists: countries who positioned themselves as exporters of primary products would perpetuate their peripheral role of suppliers to the industrial countries. Depending on a few agricultural export, commodities implied binding import capacity to those export commodities' prices on the international market, exposing the country to boom-bust cycles (Williamson 2005). Export-led strategies were consecrated a "commodity lottery"

(Bulmer-Thomas 2003: 14), since the agricultural sector was slower to respond to market signals. Agricultural products also have a more inelastic demand, both with respect to prices and to income. To make matters worse, by the mid-1920s, the BAP food commodities' prices plunged and remained low for the several following years.

Intellectually, the 1930s served as a basis for the emergence of Paul Rosenstein-Rodan's "big push theory" and Ragnar Nurkse's "balanced growth theory," which later became dominant paradigms for Latin American economic policymaking. With equivalent insights, both theories predicted that growth in developing economies would never be achieved through increased exports of primary commodities. They argued that development strategies should place greater emphasis on industrialization, laying the theoretical foundations for what would later be the import-substituting industrialization (ISI) model. This theoretical rejection of dependence on agricultural exports translated into economic growth strategies that relegated only marginal importance to agricultural exports, seen primarily as a source of foreign exchange for capital-scarce economies. Instead of pursuing productivity gains in the export sector, the policy orientation was to replace imports with domestic-made products. Structuralism and dependency theory (Prebisch, Cardoso and Faletto, Singer, Myrdal) cemented these economic conjectures into policy. The ISI strategy that followed from this school's prescriptions implied high-import tariffs and soft credit lines favoring industry, while low-import tariffs and price controls were imposed on agricultural products. Resources were channeled away from agriculture and into the non-farm sector. ISI's key operative principle was the idea of a "leading sector," capable of becoming the "engine of growth" (Nurkse 1962). In the context of a self-sufficient system, this sector would supply the necessary flow of capital to jumpstart the economic activity. The agricultural sector was perceived as having little and weak linkages with the rest of the economy, thus rendering it unfit to become this engine of growth. Moreover, because the process of growth demanded capital accumulation in its early stage, resources had to be reallocated away from the labor-intensive agriculture sector to the capital-intensive industrial one. Agriculture in this view was to serve simply as a resource base.

In the post-war context of increasing independence and nationalism, developing countries regarded agrarian-based societies as both economically and socially backward. This perception was congruent with the climate of ideas at that time in the social sciences, dominated by modernization theory and its evolutionary account of social process as a linear trend of structural differentiation and an increasing formal rationality of social action. Latin American rural structures were perceived as quasi-feudal, highly stratified, and essentially governed by tradition. The sector was dominated by a generally absentee, landowner elite, which concentrated wealth and resources at the expense of exploiting rural labor subject to serfdom conditions. The source of economic dynamisms was urban, and thus huge swaths of internal and international migrants flocked to the cities, where former peasants became the urban labor force that would fill the ranks of the mass political parties and labor unions (Germani 1965). Even culturally, the *zeitgeist* dictated that the farm was the past; modernity was in mechanization and heavy industry, in the chimneys of modern factories, in the industrial unionized urban worker. Throughout the region, a new socioeconomic and political blueprint consolidated the bias against agriculture. This model of growth, income distribution, and political survival inherently impinged on the agricultural sector, for the state had to be financed with agricultural rents. Once appropriated, these rents would finance the urban-based mass political parties.

By the 1960s and 1970s, the ideological consensus against agriculture began to crack in the face of the lack of sustainability of the ISI model. Export-led alternatives gained a momentum that would become the dominant paradigm in the region between the 1980s and 1990s. However, the conceptualization of the rural sector in the Latin American social sciences was not revised. Only economics challenged the assumptions and empirical evidence supporting the interpretive framework for the rural sector. Balassa (1971), Krueger (1978), and Bhagwati (1978) questioned the role of the state in agricultural trade policy, pointing out the failures of protection in terms of inefficiency and social cost. In a more open economy, the place for agriculture was again at the forefront due to its intrinsic comparative advantage. However, neither sociology nor political science carried out a re-evaluation of the assumptions about the rural sector in their explanatory models.

The agro-export model of international trade insertion resembles the one historically known to Latin America: Peruvian gold and Bolivian silver monetized the European economies from the fifteenth to the seventeenth centuries, while Brazilian coffee and Caribbean tobacco stimulated aristocrats and revolutionaries alike in the old continent (Topik et al. 2006: 25). Paraguay's prime export, cotton, was wiped out from the international markets with the creation of US surpluses in 1952.

The same happened with Argentine wheat. Brazil suffered a succession of busts of its leading commodities: the dominant position the country enjoyed in the rubber market was crushed in 1914 under the weight of more than 70,000 tons (tn) of Malaysian and Ceylonese (now Sri Lankan) production (Galeano 1970). A striking parallel can be drawn with the decline of the sugar producing *nordeste* under the competition from the Antilles. Indeed, Latin American economic history offers a humbling lesson in placing too much hope on these "salvation commodities." The mesmerizing effect of the 2006 to mid-2008 price hike has clouded the obvious increase in market instability. Although that recent spike is unprecedented in magnitude, it was not unique. There were at least two other periods of major rapid run-ups in prices occurring in 1971–1974 and 1994–1996. Though frequent price hikes and drops are to be expected in agricultural commodity markets as a result of their intrinsic high degree of volatility (Williamson et al. 2009), the impacts of that volatility in domestic markets and social conditions may be more difficult to manage.

Soybeans in the Southern Cone

Through experimental and small-scale operations, the first soybean plantations in Brazil date from 1882 and in Argentina from 1898. Expansion began much later in the twentieth century, while commercial exploitation of the crop did not take place until the 1940s. The late expansion had an economic rationale: previously underdeveloped oil processing techniques rendered lower extraction levels for soybeans vis-à-vis its edible (peanut and sunflower) and industrial (linoleum) alternatives. As oil extraction developed further and investment flowed—albeit slowly—into the sector, soybean demand began to grow. However, neither Argentina nor Brazil had big domestic markets for this product. In the case of Brazil, the pre-eminent position of the landed gentry of Minas Gerais under the *café com leite* Old Republic (1889–1930) and their powerful dairy interests precluded further development of a market for soybean products. International markets were not open, for the USA was self-sufficient soybean producers and Europe had a steady provision of soy cake from— in a surprising reversal trend compared to present date—China. Last but not the least, fish flour (mainly the Peruvian anchovy mentioned in the previous section) was a cheaper substitute to be used as fodder for cattle. It was the drastic reduction of this protein source in the 1970s that jumpstarted soybean production in the Southern Cone. In the beginning, only Brazil and Argentina entered the market. Later, Paraguay (as well as Uruguay and Bolivia) began its soybean production.[13]

After the debt crisis of the 1980s, the region underwent agricultural liberalization as part of the stabilization and structural adjustment programs. From 1985 through the mid-1990s, agricultural sectors were transformed: rural credit, producer price supports, and marketing services virtually disappeared. With the removal of regulations on prices and inflation spreading, the cost of land soared. In the same period, the return of democracy helped deactivate the longstanding "war hypothesis" between Argentina and Brazil. This bilateral relationship had the potential of driving most of the other political, strategic, and economic arrangements in the Southern Cone. Confidence-building measures were linked to schemes for economic and political integration. The positive cycle began in 1985 with the signature of the Argentina–Brazil Integration and Economics Cooperation Program (PICE) by Presidents Raúl Alfonsín of Argentina and José Sarney of Brazil. The shift in strategic geopolitical thinking gave rise to infrastructure development to connect both nations. Previously discarded plans for roads, bridges, ports, and other infrastructure projects were revived. Once perceived as a points of vulnerability in the event of armed conflict, the neoliberal moment recast them as an opportunity to increase bilateral trade.[14] During the next decade, investment started to flow back to the region, and a proportion of the new capital flows were channeled into the competitive agricultural sector. Paraguay underwent basic oilseed industrialization in the 1930s of *mbocayá*, which continued with peanut, soybean, and more recently, sesame.[15] Between 1985 and 1987, soybeans displaced cotton as Paraguay's top export product[16] and today represent more than 50 % of the country's exports. Geographically, production has flowed from east to west, from eastern departments of Itapúa, Alto Paraná and Canindeyú to Caaguazú, Caazapá and the fast growing San Pedro and Amambay. In Argentina, soybeans originated in the southern part of Santa Fe and north of Buenos Aires. The "nucleus zone" covers southeastern parts of Córdoba and southwestern Entre Ríos. In the last decade, it has expanded to the northern provinces of Santiago del Estero, Chaco, and Salta. Brazilian soybean production at first belonged to the southern states of Rio Grande do Sul (where it started), Santa Catarina, and Paraná. During the 1970s and 1980s, immigrants from other regions of the country moved into Mato Grosso and gradually consolidated this state's position as the leading producer. From the center west region (includes Mato Grosso do Sul and Goiás), the soybean frontier has been making its way into the north, toward the east and the more protected Amazonian west into southern Piauí, Maranhão, western Bahia and Rondônia to the west.

For the last thirty years, the BAP countries experienced dramatic increases in the area harvested, production, total supply, and exports of soybeans, as shown in Table 1.1.

The three countries' share in the international soybean market has grown for the three main forms in which the product is retailed: seed, oil, and meal. For the 2015/2016 harvests, the three countries combined accounted for over 76 % of world exports of soybean meal and 68 % of world exports of soybean oil. The BAP's fast growth in exports in the last two decades—comparing the 1995/1996 campaign to the 2015/2016 one—reveals dramatic increase: 285 % increase for Argentine soybean oil exports, 487 % for Paraguayan soybean meal exports, and a whopping 1621 % increase in Brazilian soybean seed exports (Table 1.2).

Table 1.1 Evolution of soybeans in Argentina, Brazil, and Paraguay from 1985/1986 to 2015/2016

		1985/ 1986	1995/ 1996	2005/ 2006	2015/ 2016	Percent change (1985–2015) (%)
Area harvested (1000 ha)	Argentina	3316	5980	15,200	19,700	494
	Brazil	9450	10,950	22,229	33,300	252
	Paraguay	550	960	2426	3400	518
Production (1000 MT)	Argentina	7300	12,480	40,500	59,000	708
	Brazil	14,100	24,150	57,000	100,000	609
	Paraguay	600	2,408	3,641	8,800	1367
Total supply (1000 MT)	Argentina	9165	18,042	56,549	90,748	890
	Brazil	19,445	32,753	74,565	119,803	516
	Paraguay	600	2408	3734	8870	1378
Exports (1000 MT)	Argentina	2541	2103	7249	11,400	349
	Brazil	1187	3458	25,911	59,500	4913
	Paraguay	475	1587	2380	4600	868

Source: Author's calculations based on USDA data

Table 1.2   World share, soybean exports of Argentina, Brazil, and Paraguay from 1985/1986 to 2015/2016

		World share of exports			
	Country	1985/1986	1995/1996	2005/2006	2015/2016
Meal	Argentina	14	27	46	49
	Brazil	32	40	25	23
	Paraguay	0.2	1.7	1.5	4.4
	USA	24	18	14	15
Oil	Argentina	20.2	33.5	57.2	50.2
	Brazil	14.4	33.7	25.2	11.4
	Paraguay	0.0	2.4	1.9	6.1
	USA	18.1	9.5	5.3	7.8
Seed	Argentina	9.8	5.3	5.3	5.3
	Brazil	4.6	17.0	17.0	17.0
	Paraguay	1.8	0.1	0.1	0.1
	USA	77.3	38.2	38.2	38.2

Source: Author's calculations based on USDA data

LINKAGES, COMMODITY CHAINS, AND THE POLITICAL ECONOMY OF AGRICULTURE

Everywhere, but especially in developing countries, agriculture plays an important role in the broader economic context. The macroeconomics of agriculture operate through transmission mechanisms or *linkages*. According to Albert Hirschman, linkages are at play when ongoing activities induce agents to take up new activities. Backward linkage effects are related to derived demand, while forward linkage effects are related to output utilization (Hirschman 1958: 100). Forward linkages are those that run from the domestic macroeconomy and the international economy to agriculture. It goes without saying that agriculture's ability to compete for resources domestically and globally is directly affected by economy-wide policies. Sectoral growth is affected by resource flows between sectors, which adjust to the relative opportunities offered by the different sectors over time. Examples of forward linkages are inflation, exchange rates, interest rates, government taxing and spending levels (fiscal policy), and international markets. Backward linkages are transmitted from the agricultural sector to the rest of the economy. Agriculture generates a series of linkages that have economy-wide effects, the most straightforward being employment, infrastructural, and technological developments. As any other sector, it competes for factors of production like labor and capital, provides raw materials for other sectors, and generates a component of national income. "Backward linkages" include the features that characterize the primary sector (like land tenure structures and prices) and define its particular role and relations with the economy.[17] This linkages framework connects the domestic with the international analytic dimensions through *commodity chain analysis* (CCA). This approach fits naturally as an element of the book's framework. Global commodity chains are defined as "networks of labor and production processes whose end result is a finished commodity" (Hopkins and Wallerstein 1994). This approach—also known as the *filière* tradition[18]—studies a given product, following it along a chain of activities from producer to the final consumer. The sum of activities involved in a certain commodity constitutes the product's chain or complex. Gereffi and Korzeniewicz (1994) emphasize the linkages and coordination between economic agents (raw material suppliers, processors, traders, wholesalers, and retailers), between providers of business services and finance and between economic agents and the regulatory framework. CCA provides an attractively different analytical standpoint from standard trade theory. The organizational aspect of international trade (Gereffi 1999) is largely ignored in neoclassical theory, which explains international trade as the aggregate of discrete transactions. CCA takes an interdependent view that focuses on the whole network of productive activities and the linkages binding them: trade is an integrated system rather than an isolated phenomenon (Raikes et al. 2000). Orthodox trade theory views trade patterns as determined by factor endowments, disregarding the issues of increasing returns to scale and imperfect competition. In contrast, CCA conceives them as a result of governance and the control those dominant players—private and public, foreign and domestic—exercise to maximize profits. More importantly, for the purpose of this book, CCA implicitly recognizes the social and political embeddedness of economic activity through the institutions that mediate individuals' interaction in the pursuit of advancement and gain. It acknowledges that the chain itself is the product of the interaction of purposeful actors rather than a "natural" process, enhancing the analysis of international production and trading relations. Cadot et al. (2004) introduce an empirically supported view of factor-market rivalry within an input–output linkages model. Francois and Woerz (2008) also note the importance of downstream linkages for political weight and effect on distributional policy issues. However, the CCA literature mostly limits power to relations happening within the chain. By incorporating political economy as the analytic basis of our model, we can "bring back in" the dimension of power wielding actors with strategic interactions among them that significantly influence the chain's form. Swinnen (2009) recognizes global patterns of agricultural distortions that cannot be explained by economic arguments, but are consistent with predictions from political economy theories.

Agricultural political economy approaches have almost exclusively focused on the evaluation of the level of efficiency of policy instruments (De Gorter and Swinnen 2002). Because of the centrality of efficiency as the organizing principle, the literature has found a natural synergy with the influential work by Downs (1957) and Olson (1965), the "new political economy" contributions of Stigler (1971), Peltzman (1976), and Becker (1983) and the public choice[19] approaches by Buchanan and Tullock (1962), Balisacan and Roumasset (1987), and Krueger (1996). Public choice analyzes politics as a market, and it is along these lines that public choice is incorporated in the analytical framework of this book. The individual preferences of citizens or "political consumers" are conditioned by economic circumstances, which include structure (endowments, income) and institutions (property rights, contract arrangements). There is a political demand, by which citizens demand political action through various forms of political support: votes and political contributions. "Consumers" organize into particular interest groups or lobbies to demand political action. On the supply side, professional politicians act guided by a certain preference-maximizing weighted objective function known as the political preference function (Bullock and Jeong 1994). This preference can be autonomous from or totally submissive to the preferences of their particular groups or parties. Subject to institutional and economic constraints, politicians maximize political returns by maximizing income transfers to their "winning coalition" (Bueno de Mesquita 2003: 51). Politicians have several competing functions, like maximizing the probability of getting re-elected in a democracy or maximizing their legitimacy if in a non-democratic regime. Politically self-interested actors have many motivations to shape productive chains: extracting rents, generating political support, aggregating business preferences, facilitating, or delegating policy implementation (Schneider 2004: 17). In this perspective, the policy-making process is like any other economic activity: agents (voters, politicians, and lobbyists) are rational, self-interested, and maximize their objective function in response to incentives, subject to power constraints.

Individual behavior and aggregate interest happen against an institutional framework that has to be present in the political economy analysis. Institutions are the product of iterated strategic interactions between the state and the economic sectors, competition within the bureaucracy and horizontal (between branches of government) or vertical (between local and national) power struggles. This analysis assumes the institutional frameworks governing the soybean chain to be "congealed tastes" (Riker 1980), or the crystallization of power relations into a structure. This question of institutional origins not only concerns this book but also the entire institutionalist approach to political economy. As Iversen (2006) states: "the more successful political economy is in explaining economic policies and outcomes with reference to institutional design, the more pressing it is to explain why one design was chosen rather than another." This assumes the outcome will be pareto-optimal, and implies that governmental choices about income distribution can be used to estimate the relative political weight of those among whom the redistribution takes place (Gardner and Johnson 2002). However, in the "political market," asymmetries of information translate into high transaction costs. As a result, *damages*[20] to the rules of the game are sometimes unknown, dispersed, and problematic to quantify (Hartle 1983), leading to rent-seeking behavior from actors attempting to secure (or block) changes in public policy or institutional arrangements that would increase (or avoid a decrease in) their income.[21]

Within the public choice paradigm, this book adopts the insights of the *collective action* model, in which the focus is on lobby groups vying for power. Olson (1990) specifically developed a framework to explain agricultural policy formation.[22] Government is modeled as a respondent to interest groups who organize themselves to carry out lobbying activities, and outcomes are determined by their ability to organize effectively, overcome free riding and organization costs. This is why a crucial question is what actors or agents are to be included in the models. Many agricultural political economy models focus on producers (farmers), consumers, and taxpayers. We have, however, preferred to innovate by incorporating multinational companies (MNCs), not only because of their commanding positions in the model of globalized agriculture, but because they have played a critical role in agricultural policy negotiations and debates. MNCs differ strongly from the farmers when considering their capital/labor ratio, their ability to organize politically, and their electoral strength. The latter are domestic, many, and dispersed, while corporations are transnational, few, and concentrated. Hence, it is easier for the MNCs than for farmers to organize. Two other advantages that corporations have is they are more capital-intensive than farms, and their shares of employment and gross domestic product (GDP) decline much slower with economic development than those of primary agriculture. Notwithstanding, as Wesz Jr. (2016) has noted, transnational power of agribusiness MNCs is unrestrained at the global scale; it depends upon the formation, maintenance, and exploration of relations of proximity, trust, and reciprocity with local rural producers, even to the extent of family and friendship linkages.

The empirical observations in the case studies validate the public choice hypotheses, since in Argentina, Brazil, and Paraguay relations between farmers and agribusiness follow what would be an expected pattern. Besides attempting to change the structure of entitlements to increase their share of aggregate transfers *among* the different sectors, there is also a struggle for transfers *within* any sector. The manner in which political and institutional conditions affect the use of resources in supporting or undermining particular policy instruments is of critical importance from a policy design perspective. Not only do the participating agents in the production chain strive for pre-eminence and control, political actors and governmental bodies also have interests and strategies. Regulation will lead to the emergence of certain hierarchies along the chain, indicative of specific political economy underpinnings. Our comparative look of the three BAP countries will reveal the outcome of the interaction between organized economic agents and national institutional structures. Their interests are not always aligned with those of the farmers, and this is why we find variation across the case studies of Argentina (*confrontation*), Brazil (*coordination*), and Paraguay (*colonization*).

Notes

1. Norman Borlaug, "Population Growth Requires Second Green Revolution," Nobel Laureates Plus interview, New Perspectives Quarterly (April 7, 2009) (http://www.digitalnpq.org/articles/nobel/353/04-07-2009/norman_borlaug).
2. United Nations, World Urbanization Prospects: The 2014 Revision, Highlights, ST/ESA/SER.A/352 (UN Department of Economic and Social Affairs, Population Division 2014).
3. World Bank, India Development Update, Report AUS5757 (Washington, DC: World Bank, Economic Policy and Poverty Team, South Asia Region,

October 2013), p. ii (www-wds.worldbank.org/external/default/
WDSContent Server/WDSP/IB/2013/10/16/000356161_20131016
171237/Rendered/PDF/AUS57570WP0P140Box0379846B00
PUBLIC0.pdf).
4. Figures correspond to poverty at the $2 per day (PPP) level. World Bank,
   "Results Profile: China Poverty Reduction" (Washington, DC: World Bank,
   March 19, 2010) (www.worldbank.org/en/news/feature/2010/03/19/
   results-profile-china-poverty-reduction); and World DataBank, Poverty and
   Inequality Database (Washington, DC: World Bank, 2015 (http://data
   bank.worldbank.org).
5. US Department of Agriculture, World Agricultural Supply and Demand
   Estimates (WASDE) 534, October 10, 2014.
6. Hymowitz (1970) has demonstrated with extensive historical research that
   Shen Nung is part of a legendary history of China derived from ethnocentric
   interpretations by Han historians. The Emperor is more a social archetype (he
   is believed to have taught the Chinese how to plow and sow grain, thus ending
   their nomadic nature and allowing them to settle) than a historical reality.
7. Unveiled by Henry Ford himself on August 13, 1941, the "Soybean Car"
   was a plastic-bodied car 1000 lbs. lighter than a steel car. The plastic panels
   were made from a "soybean fiber in a phenol resin with formaldehyde used
   in the impregnation." See "Ford Builds a Plastic Auto Body," *Modern
   Plastics*, Sep. 1941.
8. For a comprehensive list of soybean's uses, see Annex 1.
9. Still practiced in Argentina and Australia, the classic example is the early days
   of the American West, when most of the land of the Great Plains region
   belonged to the government and many cattlemen kept their herds on the
   public domain. As there were no fences, the cattle wandered widely.
   Ownership was indicated by branding cattle with hot irons that had designs
   on them unique to each owner. Ear tags were sometimes used in place of
   branding. Roundups were held twice a year; in spring, to brand the newborn
   calves, and in fall, to pick mature steers for market.
10. This reveals an important characteristic of soybean demand: small increases
    in per capita meat consumption—in the context of feedlot dominance—will
    lead to large increases in demand for feed proteins. Due to the exceptional
    position of soybean meal as a source of those feed proteins, these increases
    will translate directly into an increase in soybean demand (which will affect
    the whole soybean chain, from planting decisions, cost of inputs, area
    harvested, and retailing in the following season).
11. By 1964, Peru harvested 18 % of total world fish catch and produced 40 % of
    total fishmeal. Early warnings of depletion in north and central coasts
    appeared in the mid-1960s, but the industry moved operations to the
    southern coast. To remain competitive, fisheries overhauled fleets, which

increased production to 16MT of anchovy annually. By 1970, the FAO (UN's Food and Agriculture Organization) warned that the maximum annual sustainable yield could not exceed 9.5MT, but catch rose above 12MT in 1970 and 10MT in 1971. The following year production crumbled to 4 M, and to 1.3 M in 1973.

12. Hong Kong, Singapore, South Korea, and Taiwan.

13. The first soybean harvest registered by USDA in Brazil and Argentina occurs in 1985/1986, while Paraguay registers soybean harvests—though in minor amounts—as early as 1964/1965.

14. Former Argentine President, Cristina Fernández de Kirchner (CFK), recognized this in a speech she gave at the Casa Rosada on July 10, 2008: "*Estuvimos durante el siglo XX como hipótesis de conflicto el enfrentamiento con nuestros vecinos. Por eso, la falta de desarrollo de toda la Mesopotamia, por eso la Mesopotamia no tenía rutas, no tiene gasoductos, todavía hay en algunos puentes que cruzan a Brasil o al Paraguay lugares para colocar -se ríe el almirante Godoy que me mira- explosivos, de modo tal que si se venía el invasor, volaran los puentes*". Source: www.casarosada.gov.ar/index.php?option=com_content&task=view&id=4686.

15. The Paraguayan sesame producer association (Coprose) of the San Pedro department is taking advantage of the scale generated by soybean production to build an industrial park in Guayaibí and a flour-processing complex in Santa Rosa del Aguaray.

16. By 1985, 385,900 hectares (ha) were covered with cotton, yielding almost 159,000 tn. Those figures had dropped to 275,000 ha and 84,000 tn during the drought of 1986. By 1987, soybeans covered some 718,800 ha more than any other crop, with an annual output of 1 million tons.

17. For a detailed summary of forward and backward linkages, see Annex 1.

18. As developed by the researchers at the Institute National de la *Recherche Agronomique* (INRA) and the *Centre de Coopération Internationale en Recherche Agronomique pour le Développement* (CIRAD), France.

19. A substantial amount of the most relevant political economy of agriculture articles is found in the journal *Public Choice*.

20. *Damage* to the rules of the game is understood as rent seeking that leads to a social cost that is disproportionate in welfare terms. Again, the notion *disproportionate* can be operationalized as a transfer to a particular group from the social aggregate at a significant *full* social cost (internalizing costs such as the environmental; not always considered). Of course, the meaning of *significant* must not be left unanswered; but this has to be defined on a case-by-case analysis.

21. An immediate question arises: why would an actor oppose a welfare-enhancing change? The answer is that at the new state of affairs, such actor would be losing individually, even if welfare is raised collectively.

22. Olson applies his theory to explain why there are farm subsidies in the USA and Europe, while before the industrial revolution, farmers were actually taxed. His explanation for why developing countries tax agriculture is less convincing: the richer the country, the fewer the farmers. Hence, there are lower organization costs and less free riding. Fewer people to subsidize means the per-capita transfer increase (economic effect) is concentrated and thus more efficient, while the costs are dispersed. However, it fails to contemplate that there would also be fewer votes, reducing the *political* effect.

CHAPTER 2

A Super-Seeding Business

Abstract The international political economy structure of agriculture is currently a corporate-driven, vertically integrated system of global production. This is the result of two mutually reinforcing traits: the technological transformation into agrochemicals and genetically modified (GM) seeds and the economic globalization of grain trading. The strategic value of a unique asset—GM seeds with proprietary traits—has propelled these companies to a dominant position. The power of input suppliers in the new soybean mode of production has given them overriding influence, allowing them to appropriate a sizeable portion of the rents generated along the chain. These multinational corporate actors have exerted their power to create the institutional structure to govern the new resource (GM soybeans).

Keywords Agriculture · Agribusiness · GM seeds · Biotechnology · International political economy · Soybeans · Latin America · Monsanto · Embrapa · INTA · Roundup

Arguably, the single most important determinant for soybean expansion has been technological. The adoption of a cluster of three advances known in the literature as the soybean "technological package" (Barsky and Dávila 2008) has radically transformed the means of (agricultural) production:

Genetically modified (GM) seeds + Glyphosate + No-till or Direct seeding

© The Author(s) 2017
M. Turzi, *The Political Economy of Agricultural Booms*,
DOI 10.1007/978-3-319-45946-2_2

- GM[1] seeds were first developed by US chemical company Monsanto in the mid-1990s. In 1995, the US government approved GM soybeans resistant to its broad-spectrum herbicide—glyphosate—sold under brand name *Roundup*. Monsanto's soybeans are known as "RR," which stands for "*Roundup Ready*." Resistant to *Roundup*, the soybeans can be sprayed with the herbicide without being affected. Fumigation is done by large machines or airplanes without damaging the crop itself. RR soybeans were Monsanto's first commercial seed product resulting from biotech research and became commercially available in 1996, followed by Roundup Ready corn in 1998.

- No-till sowing establishes plants by sowing seed directly onto the site to be vegetated. It was introduced to reduce soil erosion, maintaining the value of the land over time. However, the technique also reduces labor, fuel, irrigation, and machinery costs. Less tillage improves soil quality by enhancing its carbon and water-retention capabilities, preventing compaction and structural breakdown.[2] Without tillage, crop residue is left intact in the field, decomposing and helping water infiltrate the soil, thereby limiting evaporation. This way of direct sowing has allowed yields to increase: less-eroded soils retain higher water content, and so instead of leaving fields fallow, it makes economic sense to plant another crop with or before the second harvest. This is why in the same soybean field, it is common for another crop to be planted, increasing a field's output and productivity. The cost equation of the producers under these conditions greatly affects the decision to plant soybeans. Even if each crop earns less, the total amount earned can be larger due to the fact that more crops can be produced at the same amount of time.

- Monsanto developed and patented the glyphosate molecule in the 1970s, and marketed *Roundup* from 1973. It retained exclusive rights in the USA until patent expiration in September 2000, and maintained predominant market share by switching operations overseas. With a GM mode of production that requires less plowing, weed control becomes a problem. Thus, the synergy between a production mode (direct sowing) and input of production (glyphosate) is natural and binding: instead of plowing to remove the weeds from under the earth, farmers eliminate weeds before planting by applying a non-selective herbicide: *Roundup*.

These technical developments were the result of a corporate strategy before anything else. In the 1980s, agrochemical corporations were experiencing declining profit margins and dwindling expansion opportunities as a result of increased regulations and fewer markets in which to expand. In response, they built on their existing relationships with farmers to enter into another, more promising agricultural input industry: the seed industry. In the 1990s, Monsanto positioned itself as a high-growth "life sciences" company, focused on agriculture, food ingredients, and pharmaceuticals. CEO, Robert Shapiro, pursued a vision of venturing into cutting-edge science to raise profits, adding seed and genomics to spin off the company's core business. The plan was to use the revenue generated by hugely profitable *Roundup* to finance R&D in biotechnology (biotech). The result was the GM technology and a series of GM seeds. In a fiercely competitive environment, Shapiro's R&D initiatives ensured the market position of his agrochemical products.[3] This development overturned existing products and markets, in a perfect Schumpeterian[4] logic of "creative destruction": permanent innovation as an imperative for survival through market share increase.

Monsanto had to diversify in order to avert losing its massive herbicide investment; this sunk capital had to be mobilized into more productive and profitable activities. Pelaez and Poncet (1999: 142) identify the two fundamental dilemmas the company faced:

a) How to induce agricultural producers to increase their consumption—and hence prolong the value—of Monsanto's main asset (Roundup) in the face of more stringent environment regulations?

b) How to generate brand loyalty in order to minimize the approaching market share loss derived from patent expiration?[5]

RR seeds were the answer to both questions. Quite literally, these were the seeds of a new agribusiness model of production. In the process of leveraging its technological base and innovating in biotech, Monsanto revolutionized agricultural production. Biotechnology is a disruptive technology and successive breakthroughs require the industry to radically rethink its very existence. Successful development of biotech markets came when companies flocked to the sector, mainly capitalizing on their chemical expertise and branching out into biotech. As a result, GM seeds spread worldwide like wildfire. By 2016, Argentina, Brazil, and Paraguay occupied the second, third, and seventh place in the "biotech mega-countries"

International Service for the Acquisition of Agri-biotech Applications (ISAAA) list with 44.2, 24.5, and 3.6 million hectares each of biotech crops, mostly soybeans. The global hectarage of biotech crops has increased 100-fold from 1.7 million hectares in 1996 to 179.7 million hectares in 2015.

A new agricultural market structure was thus established with new rules. First, the weight of private companies expanded as GM seeds spread, since GM expansion was the spearhead that would guarantee a steady flow of income derived from agrochemical sales. Ultimately, the big earnings for companies in the sector come from the chemical products. Secondly, profit would increasingly be derived from patents and royalties from seed sales. Indeed, by 2016 Monsanto owned over 16,000 biotechnology patents worldwide. Because now innovation had become the key element to enhance competitiveness in the agricultural sector, protection of the asset against imitation was paramount to safeguarding R&D investment returns (Teece 2000: 135). Organizational routines and business strategies continuously clustered around adding new and—due to intense competition—specific value to crops. The overriding trend is toward permanent development of complementary assets that will enable the appropriation of the benefits of innovation (Fuck et al. 2008: 225). Those complementary assets are integrated, inter-related components of a technology-intensive agricultural model, which in 2014 represented market values of $39 billion for seeds, $116 billion for agricultural equipment, $54 billion for agrochemicals, and $175 billion for fertilizer. As early as 1998, the Wall Street Journal reported that "most seed companies have either aligned themselves with or been acquired by crop-biotech juggernauts like Monsanto Co., DuPont Co., and Dow Chemical Co."[6] These companies are in the chemical business, and hence see the seed industry as a way to insure a growing market for their herbicides: as head of investor relations for Bayer CropScience, Alexander Rosar stated that the company's strategic priorities were "to drive top-line in agrochemicals" by means of "targeted cost savings through successful integrated crop platforms" and to "expand seeds and traits business by leveraging proprietary trait assets."[7]

In this concentrated corporate landscape, the so-called "Big Six" group stands out through the control of agrochemicals and GM seeds: Bayer, Monsanto, Dupont, Dow, BASF, and Syngenta. According to the ETC Group, the B6 collective 2015 sales were over US$65 billion in agrochemicals, seeds, and biotech traits. Together, they control 75 % of the global agrochemical market and 63 % of the commercial seed market and account for more than 75 % of all private sector agricultural research in seeds and chemicals. Moreover, the Consultative Group on International Agricultural Research (CGIAR) estimated the combined B6 R&D budgets were 20 times higher than R&D spending at international crop breeding institutes and 15 times higher than the US government's (USDA/ARS) crop science R&D spending. This means they determine priorities and future direction of agricultural research. Indeed, 2013 data from ETC showed that in the agrochemical market three companies have a 51 % market share: BASF 13 %, Bayer 18 % and Syngenta 20 %. Brazil and Argentina occupied in 2014 the first and eight positions in the top 10 agrochemical markets by country. In the global proprietary seed market, six companies control 62 % of the business: Bayer 3 %, Dow 4 %, Syngenta 8 %, DuPont 21 % and Monsanto 26 %. In December 2015, DuPont announced a US$130bn merger of its crop science division with Dow Chemical. In February 2016, the China National Chemical Corporation announced the buying of Swiss seed company Syngenta in a US$43bn deal. In September 2016, Bayer bought Monsanto in a US$66bn deal. If all these mergers go through regulatory hurdles, the three biggest companies that will emerge (Bayer-Monsanto, ChemChina-Syngenta, and Dow-Dupont) will sell 59–62 % of the world's patented seeds.

Because extraordinary gains can be captured if scale advantages are leveraged and barriers to entry raised, there is a powerful collective incentive for the sector to concentrate through vertical integration and to guard knowledge creation. In practice, vertical integration translates into significant barriers to entry for new seed companies. These include limited access to funding (established companies have built on cutting edge developments in order to consolidate a credit circuit[8]), lack of marketing experience,[9] an insurmountable R&D development capacity gap,[10] and the consequent difficulty to attract qualified scientists (in shortage in the biotech field). This is further accentuated in the countries under study, where capital markets are small and weak. For example, Bisang and Gutman (2005) estimate that the adoption of the soybean "technological package" demands an initial investment of at least US$100,000, "which makes vertical integration unviable for agricultural producers with less than 100 ha." Knowledge protection is implemented through intellectual property right (IPR) protection, which is increasingly important for guaranteeing rights and stimulating investment. Nevertheless, it could end up being utilized as an instrument to maintain oligopoly on knowledge creation and regulate the volume and pace of technology transfer according to corporate demands. As US Assistant Attorney General for Antitrust Christine Varney (the Justice Department's top antitrust official) acknowledged in reference to the American case: "patents have in the past been used to maintain or extend monopolies; and that is illegal." Because firms in the biotech field control technology distribution chains, they are political actors. They are agenda-setters, enforcing regulatory and enforcement capacities through governmental contacts and lobbying activities.

THE INSTITUTIONAL FRAMEWORKS

The structure of economic appropriation plays a key role in determining the strategies of the organizations involved in the research process, and ultimately in shaping the seed market. The distribution of the economic benefits from agricultural biotech is decided by the manner in which states have reacted to the global agro-biotech tide of innovation: what kind of regulatory frameworks were put in place, how does the institutional structure works, what have been the main interests of the actors, and what resources have they mobilized.

The basic design for the process regulating biotech crops worldwide has been modeled after the US system. It begins with a permit application for the experimentation and/or release into the environment of a lab developed crop. Tests are then carried out in order to assess potential impacts on the agricultural system, on the environment, and on the human health. After biological traits of the organism are examined, it is released in experimental crop form. During this five to seven year period, the crop is subject to recurring controls. Then, genetically modified organism (GMO) crops are authorized for commercialization. More importantly, the process is implemented differently and enforced in varying degrees in each of the BAP countries. Different national IPR protection structures will strengthen or weaken the actors in competition for control of this link in the soybean commodity chain by setting different incentives. These incentives will in turn condition investment decisions and ultimately the pace of innovation. Hence, the key arena is the R&D process and the main players are the state institutions governing it and the corporations attempting to shape it. It is this interaction that primarily determines the accretion and distribution of the benefits of innovation and, in turn, structures the domestic seed markets.

Regarding the biotech companies, they can be said to have four main avenues to advance their interests:

- *Legal*: To enforce IPR, seed companies have used out-of-court settlements and lawsuits primarily—but not exclusively—targeting farmers.

For example, after failing to reach an agreement with Argentine farming organizations and soy exporters, Monsanto's IPR claim conflict escalated in 2005. The company had freighters with Argentine soymeal cargo detained in the ports of Denmark, the Netherlands, England, and Spain in order to prove that they carried RR soybeans. In Italy, the same enterprise remained unsuccessful as the state denied Monsanto the right to detain freighters. Monsanto's purpose was to claim the cargoes illegal, as RR beans are registered in EU patent law. In June, Monsanto sued import enterprises Danish Lokale Andel and Cargill at the Danish High Court and the firm Cefetra at the Dutch Rechtbank's Gravenhage. The company issued a foreboding statement in which it claims: "The right to begin legal actions on the assumption of uncovering imports from Latin America of unlicensed Roundup Ready soy in countries where technology is protected by intellectual property rights."[11] Legal conflicts concerning seed patent rights have never been relevant in the agricultural market. With an increasing number of crops patentable, the international agricultural inputs market might—in the coming years—increasingly resemble the pharmaceutical market. If the comparison can serve as a proxy, then no better access to food products is to be expected; the patent-protected prices of pharmaceutical drugs have often put them out of reach of the poorest of the world.[12]

- *Economic*: More importantly, companies consolidate their position in the sector by a dense and complex web of subsidiaries and licensed distributors. "Cross-enabling agreements" create de facto niche monopolies. Sharing and mutual licensing of traits and technology combine R&D efforts and put an end to sector intellectual property (IP) litigation.[13] Agrochemical and seed companies are reinforcing market power from the top through deals and alliances that render futile the notion of competition. The boldest examples are BASF and Monsanto's U$1.2bn R&D and commercialization collaboration agreement in plant biotech, characterized by ETC Group as a "non-merger merger," the 2008 Syngenta and Monsanto settlement of all outstanding patent, antitrust, and commercial litigation relating to the two companies' global corn and soybean businesses, and DuPont and Syngenta's crop protection technology exchange of chemical substances. Market power exercised by dominant seed firms limit farmers' choices. By providing multipliers with incentives to limit access to seeds with weaker IP protections, they reduce the availability of non-GMO varieties. This reduction in quantity also increases the price, further discouraging acquisition. The situation even replicates for GMO seeds with single traits, which lose to seeds stacked with multiple traits. Even if multiple traits are not necessary, farmers are faced with no option to avoid purchasing them. This is why stacked traits increased from 51.4 million hectares in 2014 to 58.5 million hectares in 2015, or 14 % according to ISAAA 2016 data.

- *Political*: Subtle avenues include sponsorship agreements of public and research institutions, such as grants, joint projects, and prizes. A positive development as it may be in terms of knowledge sharing and cooperation, it should not come at transparency's expense. Financial leverage can be used in the sponsor's favor, especially in countries where institutions are weak. Not infrequently the companies use their unique expertise to advise in the drafting of national biosafety bills. Industry-friendly experts in key decision-making positions then rotate toward corporate jobs.[14] The problem here would be not lack of capabilities but of oversight, disregarding potential conflicts of interest while in public office. Lobbying for the development of strong IPR frameworks is probably the most vital element for the seed companies. The legal recognition of their proprietary traits over genetic material—by law or in court—is the most important enabler to a steady stream of profit. A final avenue is corruption, buying the laws or regulations needed through payments to officials, as was the case of Brazilian congressman Lupion, explained later.

- *Scientific*: Since companies hold an almost exclusive control over cutting-edge advanced genomics, leveraging biotech capabilities could compensate for legal or political uncertainty. Increasing "codification" of proprietary traits reduces room for piracy, though also for local adaptation. Genetic use restriction technology (GURT) has the potential to achieve IPR protection by itself, by means of science rather than law. V-GURT "terminator" seeds are genetically engineered to be sterile in the second generation, while T-GURT "traitor" seeds would not germinate until the crop plant is treated with a chemical activator compound sold by the biotech company. With adaptation and dissemination probabilities reduced, the balance of power in public/private relations could be tilted definitively in favor of the corporate sector.

Before going into the details and specificities of each country, a graphic outline of the internal workings of the seed circuit or process of production can be outlined; see Fig. 2.1.

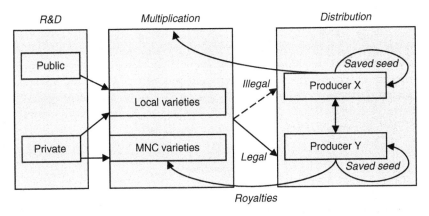

Fig. 2.1 The seed production circuit
Source: Author

THE POLITICAL ECONOMY OF SEEDS

When international chemical companies pushed forward a global biotech revolution, they created a new model of agricultural production: capital and technology—not labor—intensive. Knowledge in the biotech field was the main asset, and hence the struggle was for knowledge creation and protection through IPR enforcement. MNCs in the biotech field controlled technology distribution chains, and they exerted a political role by attempting to influence the design of the regulatory framework, lobbying for the enforcement of property rights or having "friendly regulators." The results varied in each national case: the more dispersed Brazilian power system gave rise to a decentralized governing structure, devolving power to the subnational states units and through mobilization of local producers. This led to coordination with the private sector R&D. Embrapa consolidated as a top-tier agricultural public institution, but at the same time continued to operate under profit-driven guiding principles. Argentina's research institution, National Agricultural Technology Institute (INTA) did not face incentives to develop new genetic varieties, for its climate conditions allowed direct import of foreign varieties and its institutional framework did not create incentives for profit. Its institutional structure was strong and its political mobilization was even stronger. Argentina is the only case in which there was an open *confrontation* that challenged the role of multinational seeding companies.

The impugnation of IPR debate was along rights. Finally, Paraguay's institutional structure was overwhelmed by fragile initial conditions, collective action insurmountable hurdles, and overwhelming power from the global corporate actors. Institutionally weak, it could not develop indigenous seed capacity or even regulate the power and actions of MNCs. As a result, Paraguay's structure has been *colonized*, relegated to a dependent position in the global value chain through the loss of potential benefits from soybean biotech research and development to foreign actors.

Argentina

The Argentine seed industry has been one of the cornerstones of the country's agricultural development. Scientific genetic improvement can be traced back to mid-twentieth century, with the industry organized around the activities carried out by the INTA in the public sector, and a group of local firms such as Buck and Klein, Morgan, and the subsidiaries of MNCs like Cargill, Asgrow, and NK-Nidera (Gutiérrez, p.196). In the soybean seed market, however, INTA has far less power of intervention and articulation than Brazilian counterpart, Agricultural Research Corporation (Embrapa), which was purposefully designed and institutionally sustained to be the key technological development agency underpinning the seed industry. This arrangement led to a competitive relationship with the seed companies in creating and controlling genetic material. As a result, not only is the Brazilian seed segment less concentrated, but state institutions also have a vested interest in enforcing IPR, which in practice means less lenience with the illegal seed circuit. Argentina does not have this kind of public R&D institution for soybeans. INTA has weaker capabilities because it was never able to build advantages over the MNCs as Embrapa did. Import and adaptation of technologies developed abroad was much easier in Argentina than in Brazil, where Embrapa developed a unique expertise in tropical agriculture. In consequence, Argentina held no incentives to create synergies or spillovers with local agents, or to partner with the private sector to extend the range of products or to support small and medium seed producers organized around local foundations. With the no significant public role in the soybean seed business, control was handed to the private sector and market concentration is higher. However, in Argentina, a general lack of enforcement regarding IPR and widespread circulation—both production and distribution—of illegal seed[15] results in a situation characterized by de facto transfer of the benefits of R&D from the developer to the user.

Argentine seed law 20.247 dates from the early 1970s, though enforcement began only in the late 1980s as a result of action by wheat breeders. The law provides two kinds of plant variety registration, via the National Registry of Cultivars[16] (RNC) and the National Registry of Cultivar Property (RNPC). Regulatory Decrees 2.183/91 and 2.817/91, which set up the National Seed Institute (INASE), supplemented the law. Resolution 124/91 organized the biosafety regulatory system around the National Advisory Committee on Agricultural Biosafety (CONABIA), within the Secretary of Agriculture, Livestock, Fisheries and Food (SAGPyA[17]). An advisory agency, CONABIA scientifically assesses the potential impact of the introduction of GMOs and reviews applications for field tests, supervising release of new species. However, CONABIA's weak institutionalization affects policy decisiveness. Without the competence to fix penalties for non-compliance, policy quality is also diminished. Commitments undertaken by Argentina as a member of the Convention on Biological Diversity, which requires that biosafety regulatory systems be ruled by laws, somewhat compensate for this state of affairs. During 2001, SAGPyA actively cooperated with members of the Argentine Congress in the drafting of a law on biosafety. With the crisis that wrecked the country in 2001, the draft was never brought to the floor and there is no evidence that it will be in the near future.

Argentina was the first among the BAP countries to approve RR production in 1996, and from there it was introduced illegally into Paraguay and Brazil, countries that at that time had a ban on GM crops. The loosely controlled tri-border area was used as a launching platform to introduce the seeds into Brazil and Paraguay. Black market seeds are known as "white bag" (*bolsa blanca*) for the white, unlabeled sacks in which they are stashed. Monsanto had an agreement with Asgrow in the USA for access to the RR gene. Thus, Asgrow Argentina had the right to use the gene in its registered varieties.[18] When Nidera acquired Asgrow Argentina, it gained access to the gene and widely disseminated it. The corporate strategy is the same Monsanto has used in other places of the world with other biotech crops: introduce the seed and then demand property rights based on patent law. The gambit did not pay in the Argentine case. Despite the absence of definitive evidence as to Monsanto's role, the spread of RR seeds was so instrumental to the company's objective that—given the evidence gathered during research—it is a reasonable assumption that Monsanto at least turned a blind eye to the process. When Monsanto tried to patent the gene in the country, it could not do so because the gene had already been "released." Moreover, the patent law in Argentina did not cover plants, and National Seed Law allowed farmers to save seed.

Through private settlements that explicitly recognized ownership over this patent and stipulated the royalties to be paid, Monsanto licensed the RR gene to other companies. However, institutional conditions were not granted for a biotech company to either charge a "technology fee" or restrict the use of the seed to farmers. When the company threatened in January 2004 to withdraw from the soy business in Argentina and to halt all in-country R&D programs, SAGPyA published a legal draft to initiate a "technology compensation fund." Producers challenged this "masked farmer tax" with support from the oil industry and the office of state revenues (AFIP), and the proposal was dropped.

For the eight years after 2007, Argentina laxly enforced system of "extended royalties" for soybeans, even though it is directly against the provisions of Seed Law 20.247[19] regarding free use of saved seed. Extended royalties mean that payments are in force every time that the producer (farmer) multiplies seed. Upon purchasing original seed, current price is shown on the company's list and a special emblem is stamped on the invoice mentioning which system the purchased variety falls under. Before the next crop year, the producer must make a sworn statement attesting to the amount of seed saved for planting, and the seed company will then emit a debit note for royalties. However, and because of the way in which GM soy expanded, estimates put the legal Argentine seed market at between 20 % and 35 % of the total, the rest being divided into saved seeds and the illegal market. Nidera leads the certified soybean seed market with 48 %, followed by Don Mario (29 %). Among the minor top 5 players are La Tijereta (5 %), Santa Rosa (4.6 %) and SPS—acquired by Syngenta in 2008—with 2.5 %. In this context, seed companies found it difficult in practice to enforce their IP rights to protected soybean varieties.[20]

Almost all legal GM soybean seed in Argentina is sold by the US company, which account for 50 % of the total grain production of the country. The Argentine seed market is more concentrated and private sector-driven than the Brazilian, which is more diversified and with stronger public sector participation. The evolution of soybean seed development illustrates my claim about the power of international actors wielding their power to consolidate corporate strategies. GM soybean seed developments in Argentina are performed exclusively by the private sector.

INTA's participation in the soy seed market is very limited, circumscribed to crop management, sowing techniques, and disease control. Advances achieved by soy growers in recent years are due to the diffusion of herbicide-resistant transgenic varieties and mass adoption of minimum tillage. The Institute encourages this process by "supplying an integrated crop management package," but is not a player in plant breeding like Brazilian counterpart Embrapa.

After failing to collect royalties on its first-generation Roundup Ready technology in Argentina, Monsanto started in 2011 to build a private IP regime specific to a second generation of GM varieties of soybeans. It established a private system to collect royalties on its second-generation technology. The company signed signing contracts with rural producers that wanted to have access to the new GM varieties. The system established private royalty collection. This implied rural producers would have to pay royalties on saved seeds. For Argentine Agrarian Federation (FAA), this was a direct infringement of the law of seeds. In the 2013/2014 campaign, Monsanto released the *Intacta RR2 PRO*[21] technology, which stacks insect plus glyphosate resistance together with a new private system of royalty collection, based on scientific testing. When crops get to port, before being loaded they are charged an extra fee per ton of soybeans testing positive for Monsanto's GM seeds. On average, US$15 per ton is charged to farmers if the technology is detected. Exporters such as Cargill, Bunge, Dreyfus, Vicentín, and ADM-Toepfer have agreed to act as inspectors in order to avoid having problems when handing over the grains, after cargo ships have been stopped by Monsanto in Europe over the last few years in order to collect royalties. Argentine farmers have accused Monsanto of imposing "private duties" via exporters to make producers pay for GM seed property rights. FAA has argued since 2008 that Monsanto had signed confidential agreements with exporters to act as retention agents on its behalf. On April 15, 2016 the incoming government of President Mauricio Macri passed a resolution giving the Agriculture Ministry exclusive control of the analysis of seeds in the country, rendering obsolete the Monsanto-funded network of laboratories set up to detect its seeds at Argentine ports to help enforce payment. On May 7, Monsanto issued a press release[22] explicitly defying the government's decision, saying: "The soybean technology royalty system would remain operational" and that company "will enforce its private contracts and intellectual property rights both inside and outside Argentina." In August 2016, the Argentine Ministry of Agroindustry submitted a draft seed bill to regulate the use of saved seed and copyright payment for genetically modified material. By September, it was being redrawn after complaints from farmers. Changes included limiting royalty payment to the first two harvests—currently argentine farmers are free from any obligation to pay for the rights to use second-generation seeds- and payment exemption for the roughly 19,000 small-scale producers.

Paraguay

GM seeds were prohibited in Paraguayan soil until the 2004/2005 season. Although the Ministry of Agriculture (MAG) received requests from international companies to carry out experiments and tests with GMOs, the authority responsible for authorizing the use and release of GM seeds is the National Service of Plant Health Quality (SENAVE). SENAVE was created in 2004 by law 2459, merging the National Seed Direction, the National Direction of Plant Protection, the National Control Office for Tobacco and Cotton, and MAG's office that is incharge of domestic and international commercialization of vegetable sub-products. In October of that year, Agriculture Minister Antonio Ibáñez Aquino approved the introduction of GM seeds by Resolution No. 1691. As in the case of Brazil, since the late 1990s, RR soybeans had been smuggled from Argentina into Paraguay. Once the cultivation had reached a large scale, Monsanto started to demand compensation for the use of RR technology, as it had been doing in Argentina and Brazil.

The Biosecurity Commission (COMBIO)—modeled after Brazilian counterpart CTNBio—was created by Decree 14.841 in 1997. Dependent on MAG, COMBIO was never regulated, and Decree 14.841 remained the one insufficient institutional reference governing GMO-related activity. COMBIO itself participated in drafting Decree 12.706 of August 2008, which replaced the former and changed COMBIO's name to Agricultural and Forestry Biosecurity Commission (CBAF). Roles and functions of SENAVE and COMBIO were never defined, and so the MAG is currently under process of reassigning competencies. CBAF's core functions will remain: analysis, advice, and approval of all issues concerning GMO research and experimentation, GMO evaluation, introduction and release authorization, biosecurity norms establishment, monitoring, and enforcement. CBAF will have representatives from the Environment Secretary, the Ministry of Industry and Commerce, MAG, Health Ministry, and the National University of Asunción. This inter-institutional nature is complemented with civil society and private sector representation. All ministries involved would have to jointly enforce directives, although bureaucratic interests widely vary. Without the MAG pushing forward the initiative, Paraguayan NGOs[23] and Environment Ministry[24] were its only supporters. The most comprehensive UN-sponsored study states: "There is not a defined national policy on the use of biotechnology nor on the biotechnology security. Even though some institutional initiatives in certain sectors, started with help from international organizations, appeared in the country, a national policy formulation process on biotechnology, has not yet been approved by consensus."[25] This regulatory gap is a severe state of affairs for a country where GM soybeans account for 31 % of its total cultivated land, not only with respect to the potential health or environmental consequences, but also in terms of an opportunity cost of capturing the benefits of national innovation policies.

Law on Seeds and Protection of Crops 385/94—regulated six years later by Decree 7797/00—established the creation of the National Regime of Protected Crops, providing precautionary measures to the farmer and the researcher in their tasks through the regulation of the "farmer's privilege" (which allows the farmer to save and reuse seeds and seedlings from protected varieties for the next season) and the "privilege of the researcher," which allows breeders to use protected varieties as sources of a third variety. These regulations attempt to stimulate long-term research investments and were complemented with the inclusion of biological material in patent law 1630/2000. Moreover, Paraguay has no legal provision for either traceability or labeling systems, nor have they been included under any proposed law. More importantly, there is no established policy on stacked genes.[26] Provisions and omissions of the legal framework regarding GM seed development stimulate private—and only private—investment. Public investment for national biotech developments is not encouraged; indeed Paraguay lacks a national agricultural research institute. Devoid of strong public sector research institutions and in the presence of a weak private sector to compete against multinational corporations, Paraguay has been losing out on the opportunities to capture value in this segment of its soybean chain. This situation is in line with the interest of corporations, who have located their research activities to Brazil, where they can capitalize on a strong public institution that is at the same time receptive to joint development and agreements with the private sector. From the first approval for the planting and marketing of GM seeds—which introduced four RR soybean varieties—the number of approved GM seed varieties has expanded to ten. These licenses belong to Monsanto (4 varieties), Dutch/Argentine Nidera (3 varieties), and Brazilian COODETEC (4 varieties).

The absence of native GM seed development is indicative of a poor S&T structure. The *colonization* of the Paraguayan institutional structure regulation IP in seeds is in part due to the weak capacity of the Paraguayan state in IP and in agricultural R&D. According to the latest Agricultural Science and Technology Indicators, the total agricultural R&D spending as a share of agricultural GDP is 0.26 % ($27 million), less than what spent in 2001. Neighbors, Brazil and Argentina, are at 1.82 % ($2.7bn) and 1.29 % ($732 million), respectively. The country still relies on technologies from corporations or neighboring country competitors, and there are no institutional incentives or material capacity to develop a technological treadmill. Only 35 % of the US$30 million USD in royalties paid annually by Paraguayan soy growers stay in the country through local breeding companies and INBIO (*Instituto de Biotecnología Agrícola*).

Institutional weakness[27] has political sources. Between 2003 and 2007, Paraguay had five different ministers of agriculture. Needless to say, this situation severely hampered political decision-making, bureaucratic coherence, and policy stability. Despite the fact that agricultural products represent 54 % of total Paraguayan exports, only in 2008 a project to create a national-level R&D agency—the Paraguayan Institute of Agrarian Technology (IPTA)—was presented to Congress (3788/08). Having a strong national agricultural research institute is of key importance in the context of the centrality of the role of the technological component for the soybean chain. In August 2009, President Fernando Lugo vetoed the bill (Decree 2720/09), in response to MAG's attempt to exert more technical control over the agency. Minister Cardozo had promised to write up a new bill "agreed upon by producer organizations and farmer unions in no more than 15 days, having within a month a newly created IPTA."[28] Paraguay is in the worst scenario relative to its neighbors. It lacks a privately owned—yet concentrated—seed market like Argentina, and at the same time, institutional weaknesses have prevented it from developing public sector participation. The result is a dependent position and the transfer of rents from the Paraguayan seed segment of the soybean chain to multinational seed companies or producers from Argentina and Brazil.

Paraguayan agricultural lobby groups under the Farmer's Union Syndicate (UGP) agreed in March 2005 to pay royalties to Monsanto's Paraguayan branch for the use of GMO soybeans retroactive to the 2004/2005 crop year. Organizations involved include: the Paraguayan Chamber of Exporters and Traders of Grains and Oilseeds (*Cámara Paraguaya de Exportadores y Comercializadores de Cereales y Oleaginosas*, CAPECO) representing soybean growers, the Association of Soy Growers of Paraguay (*Asociación de Productores de Soja, Oleaginosas y Cereales del Paraguay*, APS), the Agricultural Coordination of Paraguay (*Coordinadora Agrícola del Paraguay, CAP*), and the Federation of Cooperatives of Production (FECOPROD), gathering cooperatives of rural producers and accounting for over half of the country's agricultural production. Since then, Monsanto and the farmers have agreed upon the price based on the yields of the last campaign. The price is negotiated between the provider of the technology (in this case, Monsanto) and the user (the farmer), informing the government once the price is set. This royalty collection system has only been negotiated for soybeans. The system used to remunerate inventors for their technology is similar in structure to the grain program implemented in southern Brazil, designed by grower associations, grain handlers, technology providers, and seed companies. In November 2010, amidst discontent on the part of soy growers and local seed companies, the system of royalty collection was changed. Option is given to rural producers to pay when they purchase seeds, against a certificate to exempt them from payment of royalties on the sale of grain (proportionally to the amount of certified seeds purchased). Opposition to Monsanto escalated in 2012 with the news of judicial victories by Brazilian farmers. In March 2013, Monsanto offered Paraguayan soy growers a waiver on royalties on RR soybeans starting in 2014 as a way of encouraging an "orderly transition" to the second-generation GM varieties.

The weak mobilization of soybean associations made the sector vulnerable to corporate cooptation, what is in this book classified as *colonization*. The Paraguayan case illustrates a failure of the preference aggregation and interest articulation. For example, APS is a member of UGP, one of the most important business associations in Paraguay. However, UGP also includes APROSEMP (seed companies) and CAPECO (exporters). Rural producers' organizations (FECOPROD, UNICOOP, and CAP) were unable to mobilize independently from organizations of seed companies and exporters (APROSEMP, CAPECO). It is impossible to perform Olson's function of internalizing externalities in such a diverging interest scenario. Barriers to coordinated action were too high. APROSEMP and CAPECO have transnational seed companies among their members. Monsanto was a small, focused group, with a clear business orientation. With greater organizational and financial capacity, this led it to a more successful group to exert political pressure. This interwoven and overlapping structure of political representation made it possible for a corporate interest to prevail, to the extent of shaping (*colonizing*) the institutional structure.

Brazil

Federal Law 8.974/95 first established the norms regulating GMOs, and in 1998 Monsanto's RR seed was approved. However, the Brazilian consumer rights association (IDEC) and Greenpeace filed an injunction challenging the legitimacy of government biosafety policy and questioning the National Technical Commission of Biosafety's (CTNBio) scientific authority. IDEC challenged the government's claim that an environmental impact assessment was not necessary because Monsanto had testified that the GM variety was biochemically identical to conventional ones. IDEC argued CTNBio did not have the legal basis to waive the assessment, since it is required by the 1988 Brazilian Constitution. IDEC even asserted the decree which established CTNBio was unconstitutional. After 21 months of appeals by Monsanto, federal judge Antonio Souza Prudente declared in June 2000 that clause xiv of article 2 was indeed unconstitutional. After the federal court in Brasília upheld the ruling, a non-defined moratorium on commercial planting of GM crops was enforced. Months later, in September 2003, the European Parliament adopted two regulations imposing traceability and labeling of GMOs in food products for human and animal consumption. Faced with the potentially massive losses that would accrue to the Brazilian agricultural sector, President Lula da Silva signed a decree authorizing the temporary sale of RR soy for the 2003 harvest and planting and sale for the 2004 season (provisional measure MP 113), making the courts' decision invalid. From then on, the government's strategy was to issue successive provisional measures *Medidas Provisórias* (131 and 223, PLV 67/04). Ultimately, the issue would be decided in the battle for the Biosafety Law and for CTNBio's control.

Congress approved the Biosafety Bill (N. 11.105) in March 2005. This bill replaced the previous legal framework from 1995, under which agricultural biotech was first developed in Brazil. Signed by Lula on March 24, the law includes provisions for stem cell research. On November 23, the president signed Decree N. 5591 implementing the law, thus establishing the two pillars of the Brazilian regulatory framework for agricultural biotech:

- The National Biosafety Council (CNBS) falls under the Office of the President and is responsible for the formulation and implementation of the national biosafety policy. Presided by the Chief of Staff of the

Office of the President, 11 cabinet ministers comprise CNBS, with a minimum quorum of six needed to approve any relevant issue.

- CTNBio is under the Ministry of Science and Technology and not under the Ministry of Agriculture, as is the case with Argentina and Paraguay. Although under the current law CNBS is supposed to handle all political, social, and economic issues that might impact regulatory decisions related to agricultural biotech, it is CTNBio which has been the object of corporate lobby and anti-biotech groups (these latter ones challenging its existence on constitutional grounds). In spite of formal attributions indicating otherwise, actors identify the real locus of power as being in the CTNBio.

Environment minister Marina Silva was the main political power behind the approval of the Biosafety Project by the Deputies' Chamber in March 2004. The battle for the law gave rise to the formation of two coalitions around the GM issue: the one in favor was composed of scientists, representatives of biotech companies, farmer's associations,[29] and representatives of the government like Agriculture Minister Roberto Rodrigues. IDEC and Greenpeace spearheaded the opposition coalition, but public appeal was weak and, in consequence, support was dislocated. The opposition included a variety of interests and objectives that—overall—weakened the unity of purpose and action. In a perfect example of Olson's collective action, the smaller, more coherently integrated group succeeded. Minister Marina Silva's coalition suffered a complete defeat at the Senate with the approval of a modified version of the bill, which concentrated decision power on CTNBio. The new composition of CTNBio included several environmentalists opposed to biotech, leading to frequent deadlock with regards to decisions on research and commercial approvals of new products. However, as power shifted, this situation gradually changed. CTNBio's course illustrates a shift in the balance of power indicative of the consolidation of the agribusiness model. In March 2007, the Commission's quorum requirements for votes on GM products were lowered. In 2008 alone, CTNBio approved 7 of the total 12 licenses since it began work. Increasingly isolated in the government for her views on infrastructure projects, biofuels, and GM crops, Silva left the Environment Ministry in May 2008.[30] She cited "growing resistance found by our team in important sectors of the government and society" as the reason for her resignation. However, M. Silva's model of growth was opposed to Lula's, who has put all his political capital behind a decisive championing of the Growth Acceleration Program (PAC).

This program is currently seen as the cornerstone to unlock the country's economic potential and to boost its growth rate, accelerating economic activity through increased public sector investment, particularly in infrastructure and social programs.

Regarding the institutions governing the Brazilian seed market, the National Plant Variety Protection Service (SNPC) was set up in 1997 to provide support for activities involving plant variety protection. The Seeds and Plants National System Law of 2003 created the National System of Seeds and Plants that foresees the identification and quality of multiplication materials and of vegetal reproduction used and marketed in Brazil. The law creates two registration authorities: the National Registry of Seeds and Plants (RENASEM) and the National Registry of Plant Varieties (RNC). Seed law number 10.711/2003 explicitly prohibits purchasing seeds from a producer or trader not registered in RENASEM, or growing seeds or plants using not registered in the RNC. By May 2015, out of the 32,542 total registrations of RNC only 6 % (1,954 crops, 811 of them) were soybeans. The remaining 30,588 were classified as conventional cultivars and species. GM crops established IP rights for plant varieties, thus favoring the creation of an association of breeders and seed companies, Braspov, with oversight and enforcement role. In 2004 Braspov joined ABRASEM, the national association of seed and seedling producers and leading representative entity for the segment engaged in research, development, multiplication, and commercialization. According to ABRASEM's President Iwao Miyamoto, 80 % of soy GM sales are made by Brazilian companies: Cooperativa Central de Pesquisa Agrícola (Coodetec) produces 50 %, Embrapa 31 %, and the remaining volume is distributed between Pioneer (DuPont), Monsoy (Monsanto), Syngenta and Fundação Centro de Experimentação e Pesquisa (Fundacep), and the Mato Grosso Foundation (Fundação MT).[31] They all develop their own research programs, alone or in cooperation with Embrapa and other private or public research organizations.[32] The private sector in Brazil has taken more interest in developing soybean varieties as a result of law No. 9456, which spawned the growth of the market for transgenic seeds. Embrapa and Monsanto have the most soybean cultivars[33] protected under the SNPC. However, Brazil is the only country in the world in which the state agency holds more IPRs than Monsanto: 34 % against 19 % of the total pool (Fuck et al 2008: p.229). When the Brazilian market opened up for the GM soybean, Embrapa lost its preferential position (Fuck et al. speak of "hegemony") in the soy seed market. The change in the property regime limited Embrapa's public function: the commercial criterion for the development of new varieties was bolstered and low-price distribution reduced. Indeed, the institution's self-perception changed. The seed portfolio was increasingly viewed as an asset to be protected, so Embrapa—unlike Argentine INTA—had a stake in supporting IPR enforcement. The portfolio could now be exploited commercially as well, so Embrapa charged royalties and launched agreements with MNCs and local foundations, preserving thus its public function and maintaining ownership and control of its seed traits. This articulation made it possible to adapt soy to tropical conditions, enabling it to be grown in various regions of Brazil and expanding the agricultural frontier.[34] At first, companies obtained regulation of royalty payments by the stipulation that the seeding company had to present a fiscal receipt for the sale of those seeds for which it intended to collect payment. But under the reality of widespread seed piracy, collection was rendered impracticable, and so the strategy was adjusted. Producers—mainly led by the southern states—agreed to pay a percentage (sales tax per bag) at the moment of handing over their harvest to the traders (who keep a percentage as well for taking on this service). Soy harvested had to be declared GM or be subject to testing and eventual penalties. However, the other component of the technological package, the herbicide Roundup, was not legal. Paraná congressman Abelardo Lupion pushed through a series of federal amendments that legalized glyphosate in Brazil. On May 8, 2006, the *Correio Braziliense* uncovered Lupion's corruption: he had received in return from Monsanto the Santa Rita *fazenda* for a third of its market value.[35]

When on September 15, 2009, Representative Nazareno Fonteles held public hearings at the Committee of Agriculture in the Chamber of Deputies to discuss IP agricultural issues, the Brazilian Ministry of Agriculture stated that the government had decided not to interfere in agreements that had been reached by private actors, since rural producers had agreed to pay for the use of the RR technology. The weak coordination between civil society (rural trade unions and federations of the rural sector) and politics meant institutionalization would only happen at the state and not at the federal level. This is consistent with the structure, dynamics, and historical patterns of Brazilian politics. Unlike the argentine *confrontation* example, in Brazil the pattern of *coordination* is exemplified by the fact that there is no contestation from the producers to private IP rights on seeds. In April 2012, a first-level judicial ruling on the lawsuit started by soy growers led by APROSOJA-RS determined the suspension of royalty payments on RR soybeans. According to the ruling, the only IP law that could regulate the relation between Monsanto and soy growers was the Law of Protection of Cultivars. Monsanto did not have the right to charge royalties on the total output sold by rural producers, and the latter had the right to cultivate saved seeds at no cost. The ruling included an inspection that determined patents on which Monsanto was basing its claims had expired in Brazil. Immediately following, the Federation of Agriculture and Livestock of the State of Mato Grosso (Federação da Agricultura e Pecuária de Mato Grosso, FAMATO) started another lawsuit against the company to carry out another technical examination. This one also concluded Monsanto's patents on the RR technology had expired in 2010. While FAMATO-MT was battling in court, APROSOJA-MT was explicitly declaring: "*We approve of royalty payments. However, we defend the fact that their charging should be fair and supported by Brazilian patent legislation*"[36]

Monsanto suspended royalty collection nationwide, but negotiated new IP agreements with organizations of rural producers at the state levels. By January 2013, the company had reached understandings with CNA and five rural associations at the state level. Mid-year attempts to co-opt farmers were partly effective. In exchange for a 16 % discount on royalties to be paid on the second generation of GM soybeans over the next four years, FAMATO decided dropped its lawsuit. This system eventually collapsed for being rejected by the producers associations. By 2015, the situation had changed. Justice of Rio Grande do Sul granted an injunction prohibiting Monsanto to collect 7.5 % royalties on soybean sales with *Intacta RR2* technology produced with seeds saved by farmers themselves. The lawsuit had been filed by the Association of Rio Grande do Sul Soybean Producers (Aprosoja-RS), the Federation of Agricultural Workers (FETAG) of the three southern states (Rio Grande do Sul, Santa Catarina, and Paraná) and by rural unions. *Gaúcho* producers, represented by FETAG and rural unions also have since 2009 a court dispute against payment of 2 % royalty on sales of the first-generation Monsanto soybean RR1, which already had the patent considered expired. The company had lost the action in the first instance but reversed the decision before the Court of Justice in 2014. The producers appealed, but there is no date for the retrial of the case. The ruling strengthens similar decisions adopted by the Justice of Bahia and Mato Grosso, again with scope only in those States.

Public discourse and institutional practice in Brazil does not fundamentally challenge the agribusiness model and its technological or economic basis. It is focused on distributional aspects of IP (the balance between

R&D investments, productivity gains, and the precise royalty pricing). Nor it is framed in terms of private gain vs. public interest. Different organizations and state-level associations have carried out mobilization of Brazilian soybean growers over seeds. They all diverge in the approach they take, do not coordinate their actions nationally, and articulate a public discourse focusing on short-term distributional issues. Producers are more concerned with short-term distributional implications of IP rights than with their long-term substantive nature as legal rights, a perspective that can be only brought in through national, public institution political presence.

Notes

1. The literature also uses the terms genetically modified organism (GMO) or genetically engineered organism (GEO). The notion, however, is the same: organisms whose genetic material has been altered using genetic engineering techniques known as recombinant DNA technology. DNA molecules from different sources are combined into one molecule to create a new gene. This DNA is then transferred into an organism, giving it modified or novel genes. Transgenic organisms—like RR soybeans—are organisms which have inserted DNA that originated in a different species.

2. Although direct sowing reduces soil erosion, it does not counteract the impacts caused by continuous cultivation of the same crop and the intensive use of agrochemicals as the only weed management.

3. Covers herbicides, fungicides, and insecticides.

4. Joseph A. Schumpeter coined the term "creative destruction," as the "the opening up of new markets, foreign or domestic, and the organizational development [...] illustrate the same process of industrial mutation, that incessantly revolutionizes the economic structure from within, incessantly destroying the old one, incessantly creating a new one" (1942: 82–85). Trying to understand what firms would be better positioned to innovate, he connected the ability to innovate to a company's size. Larger corporations with some degree of monopolistic power would have an advantage to innovate because of resources and scale. "Innovatory discontinuities" upset the equilibrium and generate a transitional dynamics converging to a different state of technology that will restructure the whole market in favor of those who grasped them first.

5. As a result of patent expiration, Roundup pricing was expected to fall. Monsanto took several measures to build barriers of entry: it lowered the price of glyphosate 16–23 % in 1998. It built a huge facility in Camacari, Brazil, to increase capacity by 35 % and thus dissuade competitors from committing the capital for capacity additions. Finally, it set up long-term

supply agreements with several major manufacturers and potential competitors Cheminova, Dow Chemical, Microflo/BASF, Nufarm, and Syngenta (through Zeneca and Novartis). It was able to maintain an 80 % market share in glyphosate for six years after the patent expired by tying its use to proprietary Roundup Ready seeds, even though its prices were three to four times higher than generic glyphosate.

6. "Germany's Agrevo buys Cargill seed operations," *Wall Street Journal,* September 29, 1998.

7. *Positioned for Growth* investor handout, June 20, 2008.

8. Initial biotech successes provided the capital to support ("cash cow") several seed varieties while earning revenues from seeds already in production.

9. Early successes like RR soybeans have positioned companies to market new products to growers and soybean processors who have had experience working with the previous generation seeds.

10. A simple comparison will illustrate this point. Brazil's Embrapa 2009 budget stands at US$777 M, while Argentine INTA is almost a third of this amount at US$260 M. Paraguay has a bill to create an institute (IPTA), but has not even been assigned a budget line item. On the other hand, Monsanto estimates its R&D budget at 9.5 % of its sales. Net sales for the company's fiscal year 2008 were US$11.4 bn, 9.5 % of which is US$ 1.1 bn. Even considering that Latin America accounts only for 22 % of the company's global sales and assuming a proportional relation between sales and R&D budget, the figure would be US$240 M. This is just one company and does not include collaboration agreements or joint developments.

11. "Seeds of dispute," *The Guardian,* February 22, 2006.

12. See Jeffrey Sachs, "Patents and the Poor," *Project Syndicate,* April 2001. Also Jagdish N. Bhagwati, "Patents and the Poor: Including Intellectual Property Protection in WTO Rules Has Harmed the Developing World," *CFR,* September 2002.

13. Litigation over R&D issues but not over wholesaling. This reveals companies are in drive toward increasing the volume of the market without relinquishing control. On May 18, 2009, Monsanto filed a lawsuit in federal court in St Louis against DuPont and its subsidiary, Pioneer Hi-Bred International, for unlawful use of its proprietary RR technology. Monsanto argues DuPont may not combine ("stack") its herbicide technology with any soybeans already containing Monsanto's Roundup Ready trait.

14. The "revolving door" phenomenon is by no means limited to emerging markets: former Monsanto attorney Michael Taylor was appointed FDA Deputy Commissioner for Policy (a newly created post), in July 1991. Having formerly worked on the legalization of GM bovine growth hormone, Taylor helped declare GM seeds "substantially equivalent" to non-GM seeds, hence establishing tracking and labeling unnecessary.

Former USTR Chief Agricultural Negotiator Richard Crowder was CEO of the American Seed Trade Association for the three years prior to his appointment from 1994 to 1999, as Senior VP International of DEKALB Genetics Corporation.

15. In soybeans, official Ministry of Agroindustry 2016 figures put the share at 85 % of total production, in a market in the range of $300–$450 million.

16. A cultivar is a plant variety that has been produced in cultivation by selective breeding.

17. The Secretary depended of the Economy Ministry. On October 2009, President Cristina Kirchner upgraded the Secretary to a Ministry. In December 2015, the incoming Macri administration renamed it Agroindustry Ministry.

18. Monsanto has grown its seed business lines by acquisition: starting in 1982 with Jacob Hartz Seed Co. Monsanto has purchased 15 different seeds (Asgrow Agronomics, Holden's Foundation Seeds LLC, Corn States Hybrid Service LLC, DeKalb Genetics Corp., Channel Bio Corp., Seminis Inc., NC + Hybrids, Fontanelle Hybrids, Stewart Seeds, Trelay Seeds, Stone Seeds, Specialty Hybrids and Stoneville's cotton) and biotech companies (Agracetus and Calgene).

19. The Mauricio Macri administration (2015–2019) is working on a new seed law, since 20.247 dates from 1973, when agricultural biotechnology did not exist.

20. According to GM Campaign Coordinator for Friends of the Earth Europe Helen Holder, patents have allowed the company to legally prohibit seed saving and to sue farmers that save seed. In the USA alone, Monsanto has 75 employees and an annual budget of US$10 M allocated to target around 500 farmers a year. Taking "out-of-court" settlements into account, Monsanto has collected between $85 and $160 M from farmers. *The Future of Food: Transatlantic Perspectives* International Conference, Boston University, May 9, 2009.

21. Monsanto argues Intacta RR2 Pro is a biotechnology invention that is protected in Argentina and other countries around the world by patent rights owned by Monsanto and its affiliates. In Argentina, two patents have been issued that protect Intacta RR2 Pro (Patent AR026994B1, "New constructs expressed in plants and method for expressing a DNA sequence in plants" and Patent AR010897B1, "Method for controlling infestation of a soybean plant by an insect of the family tortricidae.") Additionally, Monsanto has four patent applications pending for Intacta RR2 Pro soybean products in Argentina.

22. http://news.monsanto.com/press-release/vegetable/monsanto-discus sions-ongoing-argentinas-government-latest-soybean-innovation

23. The most visible of which are: the Asociación de Organizaciones No Gubernamentales del Paraguay, Red de Organizaciones Ambientalistas del Paraguay, Red Rural de Organizaciones Privadas de Desarrollo, Federación Amigos de la Tierra América Latina y Caribe, Red de Acción en Plaguicidas y sus Alternativas para América Latina and the Movimiento Agroecológico para Latinoamérica y el Caribe.

24. Sources inside the Ministry explained the main reason for this support is that under the Cartagena Biosafety Protocol, Paraguay was eligible to receive funding from the United Nations Environment Programme to develop a national biosafety framework.

25. Development of the national framework of security of biotechnology for Paraguay, United Nations Procurement Division, Project N. 47.054, 2007.

26. The combination of several genetic traits into one line.

27. According to Levitsky and Murillo (2005), institutional weakness should be defined negatively, as the absence of those attributes that define institutional strength. Institutions are strong when the rules that exist on paper are enforced and stable and weak when they lack one or both of these dimensions (2–3).

28. "Redactarán nuevo proyecto de ley para crear IPTA," *ABC Digital*, August 17, 2009.

29. "I like (Minister of the Environment Carlos) Minc because he will not be as radical as Marina, she is an obstacle to economic development" said in an interview Rui Prado, head of the agriculture federation of Mato Grosso.

30. Director of public policy for Greenpeace Brazil Sergio Leitao said on that occasion, in reference to the Amazon: "It is time to start praying."

31. Presentation at the Seed Association of the Americas Congress, Brasilia, September 29, 2008.

32. Examples of these are the Monsanto/Embrapa project to use conventional soybean varieties adapted to the Amazon climate and introduce the glyphosate-resistant gene and the Embrapa/BASF project to create new transgenic seeds for the warmer climates outside southern Brazil.

33. Plant variety deliberately selected because it carries specific desirable traits. For genetically modified plants, having the appropriate cultivar is directly related to propagation success.

34. As an example of the astounding market segmentation allowed by GM technology, Syngenta's NK 7074 RR seed was developed especially for the Center-West region of São Paulo and Minas Gerais, while VMax RR and Spring RR are suited to Mato Grosso do Sul.

35. Sources from the Comissão Pastoral da Terra (CPT), Regional Paraná state have confirmed Lupion was known as the *Deputado do latifúndio*.

36. www.aprosoja.com.br/noticia/aprosoja-esclarece-perguntas-e-respostas-sobre-royalties-rr

CHAPTER 3

Global Trading

Abstract Traders and processors have taken advantage of the grain trade liberalization of the last decade to leverage their position in open markets. They have concentrated supply mechanisms through the advantages derived from scale and vertical integration. Their strategies for furthering their position within the soybean chain have included infrastructure development, financial leveraging, and flexible sourcing. Further, they have drawn on their financial strengths to dictate tax structures and infrastructural developments, thus creating a pull force to rearrange the economic geography through the three countries.

Keywords Agriculture · Agribusiness · International political economy · Latin America · Soybeans · Trade · Bunge · Cargill · ADM · Dreyfus · COFCO · Nidera · Grobo

In the same way the seeding companies increased their assertiveness over the soybean chain due to exceptional technological innovations, grain-trading companies ("traders") have gained power within the soybean chain as a result of the way in which commercialization was (re)structured in the 1990s. This chapter traces the process in three parts. First, it sets the context: in the world and through the South American region, there was a withdrawal of the state after the liberalization wave of the 1980s and 1990s. Grain trading and processing were no exceptions to the general trend, which manifested in the BAP countries (Brazil, Argentina and

Paraguay) with increased privatization of agricultural markets. Second, the chapter examines the main players who benefited from this process. Liberalized markets opened the door for higher corporate ownership. Traders globalized their strategies and regionalized grain-trading operations. Powerful scale advantages generated incentives for commanding—even collusionary—control of soybean trade and vertical integration with processors, resulting in increased concentration of the sector. Finally, three aspects of the soybean-trading link are used as evidence to demonstrate how different coalitions in each of the BAP countries regulate this process: financial markets (flow of money), export duties (flow of the product abroad), and infrastructure (physical flow of the good).

The New Global Agricultural Trade

Just as the biotech revolution transformed the soybean chain not only in the physical product but also in the mode of production, the international liberalization wave of the 1990s fundamentally changed grain-trading channels and oilseed marketing structures. These changes have had an impact on the organization of the soybean chain and the relation between its links. Two simultaneous processes converged in order to reshuffle power along the chain: the withdrawal of the state in the context of domestic liberalization in Latin America, which opened up a space to be rapidly filled by the new global corporate expansion strategies of the MNCs involved in agricultural trade.

Since the 1930s, governments in South America sought to cushion the volatility effects of price swings of agricultural products on domestic food provision. Successive military and civilian governments attempted to maintain a single marketing channel for key commodities by intervening—to control or direct—the functioning of agricultural markets. In the ISI model adopted through Latin America, agricultural rents were the key input to finance industrialization. This was not only an accounting exercise but also a political imperative: the funds that kept the urban-labor-based populist coalitions together originated in the capture and redistribution of agricultural exports rents. A common instrument used was the establishment of marketing boards. Boards had a legal monopoly to purchase crops through very discretionally assigned trade licenses. The theory behind it is that price controls seek to adjust the market in order to ensure a certain allocation on the basis of social and political objectives rather than on individual preference. Institutional economics states that when four firms control 40–50% of a market, it is no longer competitive, as these dominant firms can simply signal their intention to raise prices and the other will find it in their interest to follow suit (Scherer and Ross 1990). High levels of concentration also create incentives for firms to explicitly coordinate/ conspire in order to fix prices. Economic incentives to engage in non-competitive activities would decrease by instating a state agency as the last resort buyer and seller. In the presence of collusionary behavior to artificially drive grain prices up or down, the state agency would sell or buy its stocks to stabilize the market. However, these price-fixing regimes create major incentives to cheat, since they constitute command and control measures opposed to the ubiquitous nature of markets. For policy to triumph over incentives demands extraordinary regulatory costs, convoluted—usually baroque—institutional structures, an array of auxiliary rules, and a large number of monitoring bureaucrats. Although the effectiveness of this public sector intervention varied, bureaucratic controls not only created parallel markets but also crippling opportunities for rent-seeking behavior by public officials.

The debt crisis[1] forced the Latin American governments to decrease support and review agricultural policies. Deregulation reduced state-created distortions, trade barriers were unilaterally reduced, and private financial instruments introduced. As a result, agricultural commodity markets were liberalized. By the 1980s and 1990s, the notion that government intervention introduced distortions in the market and inhibited the development of a vibrant private sector became a dominant consensus. Under this paradigm, the reform strategy was guided by central structural adjustment and market liberalization. For the South American agricultural markets, these strategies were involved: price liberalization, removal of quantitative and administrative controls on trade, market-determined interest rates, adjusted exchange rates, privatization, and contracting out of public enterprises and competences, and the abolishment of licensing arrangements. With the structural adjustment and reform programs of the 1990s, the state withdrew from direct involvement in commodity markets, and state-trading enterprises—to the extent that they existed in the BAP countries—were dismantled. In Paraguay, most direct price interventions in agricultural markets were eliminated in 1989. Argentina dissolved its National Grain Board in 1991, with presidential decree N° 2284/91. Its objective was to stabilize prices without engaging in trade, unlike the Brazilian National Supply Administration, which stabilized prices by trading alongside other enterprises. The agricultural reform process undertaken in the BAP countries was in line with privatization of state-owned enterprises, elimination of marketing boards and subsidies, and removal of export taxes (ETs), guaranteed prices, and government-owned stocks.

However, according to Murphy (1999), instead of leading to functioning and competitive agricultural markets, the retreat of the state led to equivalent market distortions, only this time due to the strengthening of multinational private companies acting with cartel behavior. Rather than creating an open market, the risk was relocated away from the state and assigned to vulnerable producers and rural workers. In his study of the Michoacán highlands in Mexico, McDonald (1999) identified the effects that the neoliberal doctrine had on agricultural markets:

> On the material level, the state is removing the socioeconomic safety net formerly comprised of such things as tariff barriers, price supports, production subsidies, and access to credit. Access to credit and other state-based resources is the inducement for those who, from the perspective of the state, respond appropriately and organize. On the level of signification, the move to organize is being done in the name of quality, efficiency, and the like. In the process, the concept of choice is valorized. However, the manipulations of behavior designed to get farmers to choose to organize and adopt new techniques of production are occurring under conditions in which the prescribed outcomes are virtually impossible for farmers to fulfill. Neoliberal ideals and on-the-ground reality exist in stark contradiction.

Mechanisms for regulating production activities were eliminated and replaced by the "competitive" pressure of foreign markets. Without these state institutions, producers have the potential to capture a better share of export prices. However, *ceteris paribus*, trading MNCs—not producers—hold the assets necessary to succeed in deregulated agricultural markets. They have insurmountable financial superiority and credit, access to multiple markets, and huge informational asymmetries that translate into cost and scale advantages. Although the sector has experienced improvement in the *generation* of efficiency and profit margins, the governance of each particular product chain will determine the *distribution* of the accrual of benefits and bearing of the costs. For soybean trading, chemical and trading companies have been the clear winners of the process in the three countries under study. At the same time, small producers are at a greater disadvantage, for the agribusiness mode of production heavily relies on gains from scale and thus pushes for the concentration of the means of production.

The dismantling of state institutions does not mean political neutrality; it is as much a political position as state intervention. Without the state, the former *locus* of political negotiation no longer exists, hampering farmers' ability to lobby or influence the market. My research suggests that it would be unsophisticated to interpret the state's decision to give up its agricultural trading role just as a move toward efficiency. Public policy can shape the soybean chain and favor specific links over others. Redefining the boundaries of the public and private spheres is not only the result of abstract economic postulates, but of the successful advancement of interests. The new scenario opened up space for private sector action. With renewed regulatory conditions, foreign direct investment (FDI) was attracted to the region. Although the vast majority was directed toward the manufacturing and services sectors, the agricultural sector also benefited from the 1990s wave of liberalization. According to FAO's *Global Agricultural Investment Dataset*, developing countries accounted for 50% of all FDI inflows to agriculture, with Asia and Latin America as the largest recipients since 1997. The countries with the highest average inflows to agriculture with the top 5 recipients for the period 1991–2013 are China, Argentina, Indonesia, Brazil, and Uruguay. China was at the same time the world's largest recipient of inflows—averaging $1 billion annually— and the world's largest foreign investor in agriculture, spending an average of $337 million per year. International funds were seeking higher rates of return that could capitalize on the emerging markets' price differentials, cost, and comparative advantages. Economic openness meant a free and faster movement of goods, capital, and technology. The opening of agricultural markets in the 1990s reshaped the governance of the soybean chain: market deregulation and trade openness implied a new framework for the agricultural sector. At the same time, increased FDI flows meant a larger share of international capital in agriculture under the logic of international complementation, reorganization of the supply chain structure, and consolidation of actors.[2] The result was the agribusiness model, in which agricultural production is industrialized, becoming capital- and technology-intensive and reducing its demand for labor. Specialization leads to a progressive disintegration of a unique agricultural sector of the economy, making product-specific "complexes" or agroindustrial chains, each combining agricultural and industrial dynamics, the more appropriate focus for analysis.

The transnationalization of markets induced corporate global expansion strategies, and the agricultural sector was no exception. These strategies were based on leveraging competitive advantages and capitalizing on technological superiority, which required a global operational network and large-scale worldwide investments. There is a mutually reinforcing dynamic between technological advancement, capital intensity, and corporate concentration. The actors best suited to succeed in an open market environment are those with the capacity to generate or leverage scale and technology advantages to improve their position within the production chain. What this meant for the soybean chain was that the Big 4 international grain traders (the so-called "ABCD"—ADM, Bunge, Cargill, and Louis Dreyfus) used their financial, logistical, organizational, and informational resources in order to buy local firms and assert their position in the chain. Two of Bunge's strategic pillars outlined in the 2008 annual investor report are "expand into complementary value chains" and "leverage unique operating model." ADM's top executives explained during a June 2009 presentation how they were "*leveraging our value chain and executing our strategy by expanding the volume and scope of our core model to leverage earnings power.*"[3] Through backward integration,[4] trading companies leverage scale advantages in transport, storage, and finance. Industry analysts all agree that margins in grain trading are comparatively thin and percentages charged on commissions are not extraordinary. For soybean trading in the BAP, there is also consensus that the market structure is competitive enough to prevent this kind of behavior. Nevertheless, experts also point to volume as the decisive factor that makes this an attractive business.

This is the reason why scale is the strategic asset to be leveraged. Profit captures and market share are obtained via integration of the supply chain through flexible sourcing from diverse locations. Source diversification gives traders more bargaining power vis-à-vis producers. Purchases can be made from an extensive web of suppliers (producers) at relatively low cost. The introduction of quality and traceability standards has reinforced their position through lists of "preferred" suppliers. Buying companies can extract favorable terms through bulk buying, playing suppliers against each other, or through delisting. A large number of farmers stand in front of a few, capital-abundant, low-cost, large traders. While farmers sell their soybeans under perfect competition conditions, traders are part of a complex, web-shaped oligopoly. Bunge's VP & General Manager for Latin America, Daniel Maldonado, emphasized this point: "*Sometimes we supply customers in Colombia or the Dominican Republic with grains from the U.S. and sometimes we can serve their needs better from South America.*

If you do not have a presence in all the major origins or the ability to manage logistics, you cannot provide this flexibility."[5]

WORLD GRAIN TRADE AND THE SOYBEAN CHAIN

The soybean chain is an example of the larger system of global grain trade, which is itself a link in the food chain. These *agrifood* chains worldwide are increasingly concentrated: the number of actors is shrinking and each one vies for control of more stages of the chain: input development (seeds and agrochemicals), production, and processing, trading, and final sale for end use. This leaves the multinational trading companies in overwhelming position vis-à-vis producers. They can no longer be accurately described as agricultural commodities traders of products that farmers independently decide to produce. They are managers of the entire agricultural (global) value chain: land ownership, input supply, insurance, technical assistance, insurance, purchasing, storage, retail, transportation, and infrastructure (Murphy et al. 2012).

Soybean trade patterns through the BAP countries are representative of the broader universe of grain trading; which itself integrates the agrifood chain. These chains are characterized by the dominance of corporate conglomerates, which determine what is produced, in which ways, and how gains are distributed along the chain. This new form of production organization raises efficiency and lowers costs through adaptation and permanent innovation. Advances also demand much higher amounts of capital, both fixed (the amount of the financial investment and probability of access to credit) and working (business and labor skills). As a result, barriers to entry from new actors are being rapidly hoisted, leading to a strong process of concentration.

In producer-driven chains—such as the soybean chain—spot markets—rather than marketing—determine pricing, timing, and direction of production. Despite an apparently dispersed international system of grain trading, multinational traders coordinate their dominant position to the point that they are in reality price setters rather than price takers in international grain markets. A recent joint report by the US Soybean Export Council and the United Soybean Board explains the articulation of the links in the chain in a very clear fashion:

> A key relationship in the soybean value chain is that developed between the seed technology companies and the major soybean processors. In the U.S.

and South America, processors control soybean purchases from growers as well as the majority of export sales of soybeans and co-products. These companies control a large share of elevators that purchase soybeans directly from farmers. If a soybean processor partners with a seed technology company to pay growers a premium for new seed-trait soybeans, the seed company can charge growers a premium for the new trait seeds. Growers recover their higher input costs for the seed from the processor when they deliver the soybeans to the processor after harvest. They often have negotiated contracts with processors who want to process these new seed-trait soybeans and secured contracts to supply food manufacturers and food service operators seeking the benefits provided by the new traits.[6]

The so-called ABCD form a relatively concentrated group that controls over 70% of the global grain market (Murphy et al. 2012; Clapp 2015). These companies have extraordinary buying and selling power due to scale. On one end of the chain, they are integrated with chemical companies—like Cargill's accord with Monsanto for its international seed division. On the other end, they supply food manufacturers and food service operators. This is by no means an issue that affects only the BAP or just developing countries in South America. In March 2010, US Federal officials expressed concern about how much control a few corporations have over food supply. The US Department of Justice (DOJ), Antitrust Division, and the USDA held their first "Agriculture and Antitrust Enforcement Issues in Our **twenty-first** Century Economy" workshop. Officials pledged to examine competition and begin a new era of antitrust enforcement to balance agricultural "after decades of industry consolidation."[7] Although it is not clear what—if any—actions ultimately will result from the five hearings which will examine competition, Attorney General Eric Holder and Agriculture Secretary Tom Vilsack called the initiative "an unprecedented act of cooperation between their agencies" and said, "it would not just be a series of lawsuits but a broad policies that would ensure big companies don't have too much sway over prices they pay farmers or charge consumers."[8]

Traders play in multiple markets and can take advantage of international price differentials. Locally, in the BAP, this translates into an ability to decide whom to buy from, with the threat of including or excluding producers from their list of "preferred buyers." Chemical companies and their private R&D structures capture the benefits of biotech development. In the same way, the gains from commercial activities in the soybean-trading segment are being appropriated by an increasingly concentrated group of international firms: Cargill, Bunge, and ADM alone control 52% of global cereals trade, including oilseed trading and crushing.[9] For the countries under study here, four companies are of special interest: Archer Daniels Midland (ADM), Cargill, Bunge, and Louis Dreyfus. These four of them combined control most of the soybean produced in the Southern Cone countries through a dense network and a large number of intra-firm operations. Their presence in South America allows them to balance global presence and thus profit from differences in short-term costs, in labor and environmental standards, tax structures, subsidies, and price. Vertical coordination strategies—consolidation of the supply base—have reduced the transaction costs, de facto locking of the rest of the links into a subordinate relation, and creating insiders and outsiders. This is why the main soybean traders are heavily invested in soybean processing. As part of the vertical integration strategy, companies are controlling other aspects of the food chain. They are not only grain buyers but also retailers and oil and meal processors, establishing a variety of alliances with other players in the chain. For example, besides being the largest soybean trader in the country, Bunge Argentina is also the number one soybean oil exporter, via an alliance with national oil refinery Aceitera General Deheza (AGD).

Through mergers, acquisitions, and more flexible arrangements like partnerships, contracts, or joint ventures, they have formed a real "cluster of firms."[10] UNDP's Human Development Report 2007/2008 called ADM, Bunge, and Cargill a "cartel." It also states that in Brazil, these companies provide farmers with seeds, fertilizers, and chemicals in return for harvested soybeans. The financial strength derived from the association between international trading and chemical companies presents a dilemma for agricultural policy. Companies provide the funds—otherwise unavailable—for the farmer to adopt the soybean "technological package" and become an agent in the new agricultural economy. The innovation contained in the package has become a technical and microeconomic tool through which chemical/seeding companies and traders/processors are increasingly controlling the conditions for production and are thus reorganizing the Southern Cone territory. Because the adoption of the package by the producer is in the company's interest, the credit conditions offered are economically attractive. But inputs for growing alternative, less profitable crops are not as cheap, as the private sector does not supply them so readily and the public sector cannot or will not finance them with the same ease. According to the report, the three companies together finance 60% of the Brazilian soybean producers, although for the state of Mato Grosso that figure climbs to 85%. The main companies involved in soybean trading are:

1. **Cargill**: The largest privately owned corporation in the USA; its 1999 acquisition of Continental's grain business reportedly gave the company 45% of the global grain trade. Cargill operates in 61 countries in businesses ranging from meat processing (Finexcor) to futures brokering, feed and fertilizer (Mosaic, which split off from Cargill in 2011) production. Headquarters in Minneapolis, MN, are responsible for sales offices in Central America, the Caribbean, and South America,[11] except Argentina and Brazil, although all destination and origination offices report to Minneapolis. Geneva-based Cargill International S.A. is responsible for Cargill's worldwide trading of grains and oilseeds and vegetable oils, handling over that exceed 30 MT a year.

    1.1. *Argentina*: Soybean-related activities include flour milling, trading of flours and premixes, grain and oilseed origination, and trading. For more than 15 consecutive years, Cargill Argentina has been the leading agribusiness exporter in the country. It owns three processing plants in Quequén, Puerto General San Martín, and Bahía Blanca and owns three terminals in Puerto Diamante, Puerto General San Martín, and Bahía Blanca.

    1.2. *Brazil:* Cargill Brazil has an integrated structure of more than 130-grain origination offices. Cargill owns four export terminals in Paranaguá (PR), Santos-Guarujá (SP), Santarém (PA),[12] and Porto Velho (RO); six soybean processing plants in Mairinque (SP), Uberlândia (MG), Ponta Grossa (PR), Três Lagoas (MS), Barreiras (BA), and eRio Verde (GO); and more than 120 soybean elevators. Cargill Brazil operates in the same way as Argentina, reporting all positions to São Paulo, which, in turn, reports Brazil's overall position to Minneapolis. Brazil also sells soybeans and products free on-board (FOB) domestically and bottled-refined soybean oil under the brands Liza™ and Veleiro®. Cargill's Trade and Structured Finance business develops financial solutions for the company, suppliers, and customers. In this line, Banco Cargill S.A., is certified for commercial and investment banking. In 2012, Cargill Brazil begins producing biodiesel from soybean oil at its $63 million plant in Tres Lagoas in the state of Mato Grosso do Sul.

1.3. *Paraguay:* Cargill's central office is located in Minga Guazú (Alto Paraná), where it also has its main processing facility: Marangatú, the largest installed capacity in the country. In addition, Cargill operates in six ports (Fénix, Gical, Pabla, Paloma, Tres Fronteras, and Triunfo) and 38 units for grain reception and storage (silos and elevators). To support its export operations, Cargill has its own shipping logistics, including a large fleet of barges and tugs, thereby contributing to the development of the Paraguay-Paraná, through its continuous use.

2. **ADM:** With headquarters in Decatur, IL, the company has offices spread through the world that report sales and commodity positions to Decatur each day, among which are Buenos Aires and São Paulo. Decatur then consolidates positions and hedges in the Chicago Board of Trade.

2.1. *Argentina:* USA and Brazil are considered as soybean origination countries. ADM's primary origination office in the country is in Buenos Aires, although sales are also carried out from Santa Fe and Bahía Blanca, offices that belong to partner, A.C. Toepfer International. Toepfer (80% controlled by ADM) operates three terminals: San Martín and Arroyo Seco in Santa Fe and Ingeniero White in Buenos Aires. Soybeans, meal, and oil are sold FOB to its Hamburg office.

2.2. *Brazil:* Organized much like Argentina, except ADM owns and operates almost all of the sales offices in this country; São Paulo, Salvador, Santos, and Paranaguá being the main ones. ADM owns or leases grain elevators in five states, including 15 in Mato Grosso, 6 in Mato Grosso do Sul, 7 in Goiás, 11 in Minas Gerais, and 3 in São Paulo.

2.3. *Paraguay:* Operating from Asunción since 1997, ADM purchased local divisions of Glencore Ltd. and Silo Amambay. Headquartered now in Minga Guazú, ADM processes about 25% of the country's grain and oilseed output. ADM operates 26 elevators along the Paraná River. ADM has built up its transportation operations by purchasing a trucking company, two river transportation companies—Naviera Chaco and America Fluvial—and building over 60 barges. At present, it operates 10 tugboats and 171 barges, owning one port facility and leasing three more.

3. **Louis Dreyfus:** French privately owned Louis Dreyfus Group (LDC) is one of the largest commodity trader in the world. Main activities consist of worldwide processing, trading, and merchandising of various agricultural and energy commodities. The company's headquarters are Louis Dreyfus *Negoce* in Paris, France and Louis Dreyfus Asia located in Singapore (where Cargill's Asia Division is also located). *Negoce* controls all grain and oilseed operations. LDC is the third largest oilseed processor in South America, with extensive oilseed crushing and refining operations in the region. It has also heavily invested in the freight business: through its *Armateurs* division, it has become a global leader in bulk transportation and logistics, with a fleet of 30 bulk ships and logistics assets (floating cranes, barges, and tugs).[13]

3.1. *Argentina:* Through subsidiary SACEIF—one of the largest soybeans exporters in the world—Louis Dreyfus Commodities owns and operates Timbúes and General Lagos crushing plants and port facilities on the Paraná river. Deep-water access to large export-bound vessels and a crushing capacity of 12,000 MT/day positions the company as one of the largest, most efficient crush plants in the world.

3.2. *Brazil:* Owns and operates five soybean crushing plants in Brazil. Subsidiary Coimbra owns and operates oilseed crushing facilities with a combined crushing capacity of over 8,000 MT/day and a combined oil refining capacity of over 600 MT/day. LDC owns and operates 48 warehouses, a network of port and storage facilities and three crushing plants, located in Ponta Grossa in Paraná; Jataí in Goiás; and Alto Araguaia in Mato Grosso. In December 2015, Louis Dreyfus Commodities—together with Cargill—won an auction to operate a grain terminal at Santos Port for twenty-five years. LDC Brasil BSL, the consortium that will operate the grain terminal, is 60% owned by Louis Dreyfus Commodities and 40% by Cargill.

3.3. *Paraguay:* A latecomer to the country—it stepped in only in 2004—Louis Dreyfus operates through subsidiary LDC Paraguay S.A. The majority of its investments are in cotton, although it owns a port in Angostura, Central Department, with a storage capacity of 25,000tn and a loading capacity of

12,000 tn/day. Since 95.6% of the Paraguayan soybean is exported by river, LDC's most strategic investment in Paraguay came through LOGICO Paraguay S.A., a company that operates three convoys with 61 barges.

4. **Bunge:** Headquarter is in White Plains, NY, where the company's primary trading desk and risk management teams are also located. All soybean oil is traded from White Plains. Bunge compiles its entire soybean, soybean meal, and soybean oil positions there to facilitate the risk management team's ability to hedge these positions effectively. White Plains operates in a similar fashion as to Decatur does for ADM, authorizing trades based on daily global positions and as Cargill, trading out of Geneva, Bunge's primary destination soybean office. Bunge's BAP operations are handled by Bunge Global Market, the group's international marketing division that operates in more than 30 countries. South American offices are responsible for their respective soybean processing capacity and for selling FOB to Bunge's international marketing group, which controls the destination market offices.

   4.1. *Brazil*: Bunge Brazil owns and operates eight sugarcane mills in Brazil that produce sugar, ethanol, and electricity through co-generation. Bunge is the largest processor of soybeans and wheat in Brazil, a leader in the vegetable oil segment. Between factories, ports, distribution centers, mills, and silos, Bunge has 100 units, present in 17 states from all regions of Brazil and the Federal District. Bunge Alimentos has a primary soybean origination offices are located in Gaspar, Santa Catarina. Gaspar is responsible for all origination, processing, and sales of soybeans and products in Brazil. It is also responsible for monitoring and compiling all of the soybean and product positions in Brazil and sending these positions to White Plains daily. Bunge operates seven soybean terminals throughout the Brazilian territory: Ilhéus (BA), Vitória (ES), São Luiz (MA), Paranaguá (PR), Rio Grande (RS), São Francisco Do Sul (SC), and Santos (SP) and has eight processing facilities for soybean industrialization: Magalhães (BA), Luziânia (GO), Dourados (MS), Rondonopolis (MT), Uruçui (PI), Ponta Grossa (PR), Passo Fundo (RS), and Rio Grande (RS).

   4.2. *Argentina*: Bunge Argentina is the number one soybean processor and the first exporter of oilseed byproducts: meals and oils). Buenos Aires monitors all of the soybean positions in Argentina and Uruguay—origination, processing operations, and sales. The second biggest soybean exporter in Argentina, Bunge owns 6 industrial complexes (Campana, Puerto General San Martín, Ramallo, San Jeronimo Sud, T6, and Tancacha), 6 export terminals (PGSM, Guide, Quequén, Ramallo, T6 and Bahía Blanca), 8 elevators, and 9 fertilizer warehouses.

4.3. *Paraguay*: Bunge Paraguay was created in 2006, and is currently one of the major agribusiness companies of the country. It sells 850,000 tons of grains annually, ranking third in the exporters list. Bunge Paraguay integrates fertilizer distribution, marketing, reception, and storage of grains, oilseeds, and industrialization.

5. **Noble:** Headquartered in Hong Kong, Noble manages global supply chains of agricultural, industrial, and energy commodities. The group owns an extensive network of soybean storage facilities throughout the BAP, complemented with processing installations, ports, terminals, barge operations, warehouses, and elevators. Controlling the whole pipeline is an integral part of this company's strategy to leverage competitive advantage through "hands on" global presence.

5.1. *Paraguay*: With commercial offices in Asunción, Noble owns Pacu Cuá (Encarnación) barge terminal on the Paragua/Paraná waterway, has a 55,000MT barge loading facility with a 10,000MT per day load capacity.

5.2. *Argentina*: Besides operating two immense ports along the Paraná river in Argentina (Timbúes and Lima), Noble has ventured into the northwestern region of the country, building the Piquete Cabado storage facility in Salta province. In 2008, Noble—financed by the IDB and IFC—started adding vegetable oil processing capacity to its Timbúes installations in Santa Fe province.[14]

5.3. *Brazil*: With the latest dry bulk export terminal—fully operational by November 2009—for grain and sugar in Santos, South America's largest port. The new terminal allows the company to bolster its pipeline strategy and to control the product's flow from the fields to the destination market. With a cargo loading capacity of 3,000 MT/hour, it would allow a *Panamax* vessel to be fully loaded in less than 48 hours.

6. **Agri-national champions:** The new mode of production has also made possible the growth of local firms. Although Paraguay has not generated enough critical mass to develop a national trading firm, Brazilian soybean producer giant AMaggi has positioned itself successfully in the trading network due to its privileged connections in Mato Grosso. In Argentina, the development of the biggest vegetable oil industry in the world has determined that the oil refineries concentrate much of the domestic demand for the bean, essential input to sustain the country's top position in the world's soybean oil and meal markets.

6.1. *Nidera:* Founded in 1920 from the union of European grain merchants in Rotterdam, predicting it would become the gateway for European export and import trade. The company's name represented the initials of the six agricultural trade regions in which they focused their activities: Netherlands, Italy, Germany, England, Russia, and Argentina. Several senior managers emigrated after the war to Argentina, and hence the country became a major domestic originator and processor of grains and oilseeds. In late March 2014, during President Xi Jinping's first official visit to the Netherlands, COFCO bought 51 percent of Nidera. By early 2016, COFCO had suspended plans to create a global commodities trading company by integrating operations.

6.2. *Los Grobo (Argentina/Brazil):* Formerly a family-owned business, the professionalization demands of the new agribusiness model of the 1990s forced the company to reinvent itself. Straddling on the great territorial expansion made possible by the new soybean "technological package," Los Grobo incorporated technology to the production process and thus became more competitive than many of its international competitors. Since its regional expansion of early 2000s, the company operates 230,000 ha. in Argentina, Uruguay (Agronegocios del Plata), Paraguay (Tierra Roja), and Brazil (Los Grobo Brasil). Los Grobo could not compete by capitalizing on its distinctive advantages. However, the key innovation that led to their success was to detach production from land ownership. By renting instead of buying, they could achieve gains from scale and leap over the hurdle of the initial investment. Los Grobo has surpassed the 100,000 tn of soybeans only in its Brazil division.[15] In Brazil, Los Grobo is the only company offering all of the following agribusiness services: (I) input supply, including agrochemicals, fertilizers and seeds; (II) grain trading; (III) a multimodal logistic platform, partly self-owned and partly owned by third parties, which covers roads, railways, hubs, and ports.

6.3. *El Tejar (Brazil)*: Originally a livestock farming argentine company, El Tejar operated in Argentina, Uruguay (Tafilar S.A.), Brazil (O Telhar Agropecuária Ltda), Bolivia, and Paraguay. Until 2006, El Tejar exclusively rented rural properties to third parties. Originally, the business philosophy was clearly expressed by Uruguay country manager Ismael Turbán in April 2009: "*I get nothing out of owning the land, what I want is the land to be productive.*"[16] However, in 2013—as relations between farming industry and the Argentine government soured—the company shifted headquarters from Buenos Aires to São Paulo. The move to Brazil is the culmination of a shift in its strategy from leasing to owning farmland. About 65 percent of El Tejar's operational assets are in Brazil, and it is now one of the top producers of soybeans in the country.

6.4. *Aceitera General Deheza (AGD) /Molinos Rio de la Plata/ Vicentín (Argentina):* These three companies have been clustered together because Argentina has the biggest vegetable oil complex in the world (including soybean, sunflower, and peanut), thus enjoying a leading position in the international soybean oil and flour markets. AGD is owned by former Argentine senator Roberto Urquía.

6.5. *Caramuru Alimentos (Brazil):* One of the 150 largest companies in the country, it is totally owned by national capital: the Borges de Souza family. Caramuru has 4.3% of the soybean processing national market, owns 73 warehouses, located in Goiás, Mato Grosso, and Paraná and has consolidated itself in important economic segments, including animal and industrial products, consumer products, commodities, biodiesel, and logistics. In transport logistics, it has terminals in the ports of Santos and of Tubarão and railroads and the Tietê-Paraná waterway, favoring the use of multimodal[17] transport and decreasing operational costs. Caramuru has strong presence in the states of Goiás, Paraná, Mato Grosso, and São Paulo.

6.6. *AMaggi (Brazil)*: Owned by former Mato Grosso governor (2003–2010), Senator (2011–2015), and most likely Agriculture Minister in a post-Dilma Rosseff administration in 2016 Blairo Borges Maggi.[18] The company commercializes, stocks, processes, and transports (via group company Hermasa) soybean production in Mato Grosso. The trading department headquartered in Rondonópolis deals directly with the main trading centers worldwide. AMaggi has three soy crushing plants in Cuiabá (MT), Lucas do Rio Verde (MT), and Itacoatiara (AM), and has partnered with companies in the sector to build the Guarujá Bulk Terminal in the state of São Paulo. It has also invested in infrastructure development in the Amazonian waterway system, from Porto Velho terminal to Hermasa terminal in the port of Itacoatiara (AM). The Maggi group has pioneered in transforming the *cerrado* lands (Brazilian savannah) for soybean production, a model that could be replicated in bordering Paraguay.

6.7. The absence of Paraguayan champions further evinces the dependent position of this country. Paraguay has been relegated to the position of low-cost input supplier for higher value-added activities in Brazil and Argentina. It is not in the interest of the companies dominating trading and processing—multinational, Argentine or Brazilian—that Paraguay develops its own agricultural national champions.

An analysis of recent trends indicate the ABCD dominance is petering out, losing their century-old dominance of the Southern Cone's grains export market to their Asian rivals. In 2015 an analysis of shipping data done by Reuters found that Asian trading houses, including China's state-owned COFCO, bought 45 percent of the Brazil's soybean, corn, and soybean meal exports. By comparison, the share purchased by ABCD was 37 percent. This indicated an abrupt turnaround from 2014, when the four companies accounted for 46% of Brazil's grain exports compared to 36% for Asian firms. COFCO bought 51 percent each of Dutch grains trader Nidera and of Hong Kong-based Noble Group's Noble Agri in 2014. In late 2015, COFCO said it would buy the remaining 49 percent of Noble Agri.

FINANCE AND INFRASTRUCTURE AS POLITICAL ECONOMY

In terms of political economy structures, this trading structure implies the insertion into the global economy of the BAP countries' agricultural sectors is being pressed by the strategies of foreign, private companies. Within this integration paradigm, Brazil, Argentina, and Paraguay would be specializing in accordance to global corporate strategies of production. The regionalization of production, distribution, processing, and trading networks is establishing Brazil as the leading platform for R&D in seed development and soybean export to the Asian markets. At the same time, Brazil is quickly climbing the value-added ladder and attempting to convert vegetable protein into animal protein by developing its national poultry and beef industries rather than exporting the feed for foreign cattle (26% of its total 2014 production as opposed to 4% in Argentina). Installed capacity in oil refining in Argentina is leading to concentrated investment in this sector. Paraguay is relegated to a dependent position of low-cost raw input supplier for industries that capture value in the neighboring countries. Making use of the Paraguayan waterway system, soybeans are brought into Argentina and Brazil to be processed or exported. Indeed, the best way to observe the pecking order within the regional soybean chain is by observing infrastructure networks.

As a highly competitive productive sector and an increasingly integrated geo-economic unity, the soybean chain in the BAP tends to minimize barriers to trade (seeking for credits and rejecting taxes) and bottlenecks in infrastructure. Trade flows originally followed not national borders but a "hub and spokes" model. Hubs are the main production centers and the storage network that flows into the export terminals constitutes spokes. Investments are made on the basis of current and potential trade flows, looking for the areas that support necessary standards for infrastructural services. Indeed, this process of regional integration has released the competitiveness of the soybean agricultural chain by reinforcing the complementarities to create added-value production. Notwithstanding this, evidence shows domestic national political economy coalitions are playing a decisive role as explanatory variables of the final outcomes. Governmental action and national public policies have played critical roles in regulating the assets of actors in the soybean chain. Policies are the outcome of distributional struggles. In each of the soybean sector of the BAP countries, there have been coalitions which confronted (Argentina), coordinated

(Brazil), or were colonized (Paraguay) by the organization of production dictated by global agribusiness interests. Evidence of this struggle can be found in three chokepoints to the free flow of production: financial markets (flow of money), export duties (flow of the product abroad[19]), and infrastructure (physical flow of the good). However, this added value is not being accrued to each of the countries or the individual producers. The soybean-driven agricultural boom has given rise to a two-tiered circuit of production in the BAP. A pattern of increasing dualization of the agricultural economy is emerging, with the modern agribusiness sector exhibiting enclave traits. The internationalized agribusiness model generates the demand for technological advances, human capital (managerial skills), and modern infrastructure in the form of roads, rails, and ports. However, the integration of the physical infrastructure networks lowers costs for traders, processors, and input suppliers only. The losers of the process are the landless peasants and small farmers, driven out of their land because the new agricultural production model renders their activity obsolete.

Financial Instruments

With regards to finance, in order to avoid the intrinsic high risk[20] and price volatility of agricultural trade, participants have always looked for the protection of contracts. From verbal accords and written covenants onwards, the instruments have tended to formalize the relationship and legal standing of buyers and sellers by establishing transaction prices, prescribing delivery times and conditions. The Chicago Board of Trade has standardized trading and its template[21] has been replicated in the local grain exchanges in the main BAP trading outlets: the Rosario Futures Exchange (ROFEX) and the Buenos Aires Grain Exchange in Argentina (MAtBA)[22]; the BM&FBOVESPA S.A.—Securities, Commodities and Futures Exchange[23] and the Bolsa de Cereais e Mercadorias do Estado de Mato Grosso in Brazil. There is no like organ in Paraguay, which results in the country losing the benefits of financial intermediation to outsiders. Paraguay is also more exposed to price change volatility. Producers lose out the most, since no financial institution exists to promote competitive prices through spot markets, trading, and settlement systems. There is also no guarantee of payment and title transfer. With big multinational corporations having the capacity to set prices, the absence of a grain exchange in Paraguay accentuates the weakness of producers. As transaction costs rise, concentration and information[24] asymmetries—a key element of the MNC's power—have risen dramatically. Consistent with this paper's earlier explanation, this observation is the result of two variables: first, the power exerted by multinational corporations to shape the soybean production network, which does include a role for trading out of Paraguay, and second, Paraguayan institutional fragility, which has meant a lack of resources for the state to advance its own interests or present effective resistance.

Nevertheless, after establishing a deal in early 2012 to set up a futures exchange in Uruguay—a joint venture between Argentina's commodity derivatives exchange MATBa and the Uruguayan stock exchange— Argentina's futures and options exchange ROFEX helped Paraguay stock exchange (BVPASA) establish its first futures market. Grain exchanges act as a clearinghouse, reducing the risk of default. They also guarantee that all trades will be honored. Since it takes no market positions but places itself squarely between the different parties, the clearinghouse system creates efficiency. Brokering pairs all exchange-traded purchases with sales, matching inter-temporal supply and demand. The futures contract was launched in May 2012 and—despite expectations that a soybean-related product would follow—it did not come to fruition. By late 2015, the Ministry of Agriculture and Livestock (MAG) conducted a technical feasibility studies for a soybean trading in Paraguay together with Argentina, Brazil, and Uruguay, as a means to establish reference price for the crop as an alternative to the Chicago Board of Trade (CBOT). On May 2016, World Bank resident representative in Paraguay, Dante Mossi, announced in July that the country would open its first national grain exchange, an alternative/competition to Argentine ROFEX:

Operations have two main varieties: "cash" and "futures." The first variety operates with stocks readily available and immediate means of payment, while the second operates with forward contracts (FCs). Because agricultural commodities are subject to sudden price swings, the risk of default and non-compliance is extremely high. Thus, FCs are rendered obsolete in the absence of an enforcement agency. Ownership of an FC is equal to ownership of the underlying asset. However, before an FC expires, the owner has the right to liquidate the futures position or hold the position open, demanding actual delivery of the physical soybean. In fact, during 2008, the price of soybeans reached extraordinary highs of over US$1,200/tn because the majority of the futures positions are hedge trades. These speculative positions are usually taken by big traders that have the financial strength to leverage volume advantages; or hedge fund and portfolio asset managers in search of diversification, short-term attractive rates of return or coverage. On the one hand, these activities can artificially impact soybean prices, but on the other hand, they provide liquidity. Again, it is the role of public policy to put these instruments to work for the objectives desired through a set of regulations that promotes transparency without stifling efficiency. Operations "to arrive" are a commonplace innovation in the BAP. The elevator sells upon harvest, but a time lapse is established for the elevator to set the final sale price based on the daily price in the exchanges or the arbitration chambers. This arrangement solves the storage issue and offers coverage in inflationary scenarios by acting as an index. As a result, "to arrive" operations provide futures markets with a way of operating as spot markets in practice. Representatives of all the exchanges consulted point to a key tug of war taking place in the trading link of the soybean chain. Regionalized multinational exporters buy "to arrive" and then use their extensive storage network in order to leverage the volume levels and keep the reference prices low. This way, they take advantage of the slight price differentials between the several grain exchanges, making a profit in a business with slim margins. This strategy was possible in the absence of producer-owned storage facilities. Without storage infrastructure (silos), farmers are forced to quickly sell their grain, losing the speculative gains of deciding when to sell. This situation somehow changed with the appearance of silo bags,[25] empowering the producers. With this low-cost storage method, producers have more room to decide when to sell, in which provincial exchange, and at what price. At a smaller scale, technology has allowed producers to replicate the strategy of companies with global presence, which coupled with intra-firm trade has allowed the MNCs to leverage their positions in the different world grain markets, from Mato Grosso to Rotterdam and from Rosario to Singapore.

Other financial instruments have developed over time in commodity trading. *Options* on futures convey the right, but not the obligation to buy (call), or sell (put) an FC at a specified price (strike price) during a particular period of time. Grain *swaps* are over-the-counter exchanges of cash flows—one fixed and one floating—that depend on the price of an underlying crop. Usually, only the payments are exchanged and not the principal. *Spreads* or *straddles* are the dynamic price differentials between two or more contracts, usually FCs. These contracts simultaneously buy and sell futures or options with the goal of making a profit from the changing price relation between the elements of the trade. Inflation—of the two- or even three-digit levels experienced in the Southern Cone—historically distorted relative prices and reduced the level of operations in the forward market. There is no way of accurately predicting what effective future prices will be, since the correlation between prices, exchange rates, and interest rates goes awry. Thus, anticipating future agricultural rents becomes an extremely risky wager. Unpredictable or discretionary government intervention directly affects futures market. Investors do perceive (the probability of) expropriation as a risk, but agricultural cycles are longer. Multinational trading companies see a long cycle of commodities based on demographic projections of three to five decades. Further, their infrastructure investment decisions reveal a decision to operate in the BAP long-term. For example, when the Argentine government had its face off against producers in March–June 2008, traders and processors continued their operations without siding with any of the contenders, suggesting their interests could be smoothed out inter-temporally and with the companies' regional configuration of operations.

Taxes/Duties

The Brazilian current tax system is one of the most complex[26] in the world, with more than 80 taxes and numerous standards. For Brazilian companies, calculating, accounting for, and paying taxes consumes 2,600 hours per year. The tax burden was 25% and in 2015 it reached 36% of GDP.

Brazil's soybean industry has operated with far less direct and indirect government intervention. Kandir Law of 1996 zeroed taxes on foreign sales of grains. Nevertheless, the ICMS (*Imposto sobre Circulacao de Mercadorias e Servicos*), a value-added tax imposed on the movement of all goods (including soybeans and products) remains. ICMS tax rates are set by the federal government at uniform rates ranging from 5 to 13%, depending on the product and whether it is sold within the State, to another State, or exported. For some commodities, the rate also varies depending on the degree of processing. State governments collect taxes; so cost has tended to vary by State and by product. ICMS raises the cost of moving agricultural commodities. The ICMS tax continues to cause distortions in the domestic crushing industry. Crushers must pay the ICMS when they buy soybeans from other States, then recover (at a later date with a rebate) the ICMS paid on soybeans if the resulting product is exported. However, the ICMS recovery system does not appear to be functioning well. As a result, the inter-state ICMS tax system has encouraged Brazilian investments in soybean farms in Paraguay and Bolivia. Soybeans imported into Brazil are not charged the ICMS if the products are re-exported.

In Mato Grosso do Sul, a state decree (No. 11.803) was created in 2005 as a special system of supervision and control of export of soybeans and corn, taxing the ICMS 12% of foreign sales. At the national level, in 2014, Congress had struck down on a proposal to impose a social security tax (PIS/Cofin) on the internal soybean market. Then Agriculture Minister Neri Geller said President Dilma Rousseff's government did not support additional taxes on the farm sector. So the issue resurfaced again in November 2015 in Mato Grosso, Tocantins and Goiás states. In January 2016, the state of Goiás—the fourth largest producer of soybeans in the country—imposed a decree (No. 8.548) limiting exemption for exports of corn and soybeans purchased in the state. As a result, 30% of soybeans produced in the state and sold abroad will be taxed at a 17% rate, based on the ICMS. In the case of processing industry, 40% of what they buy and export will be taxed. But the argument from the state government is different: as Goiás has exported so much grain, the crushing industry ends up having to import raw material from other states to meet the demand and thus pay ICMS to other states. In this way, by limiting exports through the new tax, revenue is now being accrued according to Economy secretary, Ana Carla Abraão Costa. The Secretary also explained that "the new tax was not an ET but an export quota that the state always had and that they were only strengthening the penalties. There was a lack of soybeans in the state, and the industry purchased from Mato Grosso and other regions".[27]

On October 9, 2013 Paraguay's lower house of Congress gave final legislative approval to a bill imposing a 10 percent tax on soybean exports. A week later, President Cartes vetoed the law, despite an expected income of $300 million per year in government revenue. Paraguay eliminated in 2014 the ET on soybeans and replaced it with a value-added tax of 5 percent on the sale of all agricultural products. Another introduced change was that, if an agricultural product is processed, a 2.5 percent rebate is offered to exporters, aiming to incentivize high value-added exports through industrial processing of primary products. By late April 2016, two tax projects on soybean exports were not addressed in the Senate due to lack of quorum, after Senators left their seats as the issue came to the floor. Senators Fernando Silva Facetti and Blanca Lilac

Table 3.1   Tax rates for soybeans, *Retenciones móviles* (2008)

International price of soybean (US$/tn)	Export tax rate (%)
0–200	23.5
201–300	23.5–28
301–400	28–36
401–500	36–43
501–600	43–49
>600	>49 (reportedly limited at 50–52)

Source: Author's calculation based on Economy Ministry Resolution 125/2008

Mignarro from the Liberal party introduced one of the projects. It proposed a rate of 10–15 percent for natural or legal persons exporting soybeans. The other sought to implement a tax of 15% export unprocessed grain (soybean, corn, and wheat) and was introduced by Senators Carlos Filizzola, Sixto Pereira, Hugo Richer, Fernando Lugo, and Esperanza Martinez from Frente Guasú. Both projects were blocked when Senators abandoned the meeting to avoid giving the necessary quorum for the session to continue (Table 3.1).

Argentina was the only of the three BAP countries that applied ETs to soybean production. From the political economy perspective, Argentina uniquely signals the centralization of resources purported by state institutions vis-à-vis the productive sector. As detrimental it may be for production or difficult to understand for the analyst, it was nevertheless fully coherent and is in line with the interests and demands of the winning coalition, as will be explored in depth in the following chapter. After the 2002 devaluation, the government reintroduced ETs under Emergency Law 25.561 in 2002. As usual in this latitude, there is nothing more permanent than a temporary measure. No date was set for removing them, especially since they boost tax revenues. Moreover, these revenues go directly to the national government and are not shared with the provinces. This issue points to a historical distributional fight between the nation and the provinces that lays at the very origin of Argentine history. According to economist Pablo Gerchunoff, the proportion of total tax collection that goes to the provinces is only 27%, down from a 56% in 1989.[28] Indeed, the resistance to the ETs or *retenciones* ignited the debate about the shared federal tax revenue, cornerstone of the centralist Argentine presidency's power. With these discretionary resources, the Executive branch obtains support and commands discipline from governors directly, circumventing the Legislative. During the Duhalde government, Economy Ministry resolutions 11 and 35 established that exports of raw materials would pay 10% ET, specifying 13.5% for grains and oilseeds, 15% leathers, and 5% for industrialized products from agricultural origin. The soybean ET increased in the Kirchner administration from 23.5% to 27.5% in January 2007 and to 35% in November. Soybean oil and meal ET was raised from 24% to 32%. In 2008, already in the Cristina Kirchner (CFK) presidency, Economy Minister Lousteau established resolution 125/08, by which ETs became "mobile," in line with international price of a commodity (sliding tax). The government scale was:

The proposal triggered farmer protests and strikes, with a standoff that lasted from March until mid-July 2008, causing shortages, fall in fiscal revenue, and a collapse in the president's approval rate. This backlash was the most intense conflict of the entire CFK presidency—what some have called the "Buenos Aires soy party"[29] tributary revolt. On July 17, 2008 the Senate voted against the sliding tax in a tie-breaking vote decided by Vicepresident Julio Cobos. This defeat dealt a heavy blow to the Kirchners' grip on power and weakened the government. On July 21, the administrative measure that had originally imposed the sliding tax was suspended, ending the strike and maintaining the fixed 35% ET. After unleashing the most serious conflict of the second Kirchner administration, Lousteau was fired. Months later he would acknowledge the measure "*effectively brought futures markets to a standstill.*"[30] Not only did the action destroy the grains futures market, but in doing so, it killed the possibility of financing through futures sales. It introduced an extreme level of institutional volatility, which is especially damaging for agricultural trading, a contract-run business with long cycles of maturity. Whether by ignorance or by political intention, the government distressed the basic structure of agricultural markets.

On December 15, 2015, President Mauricio Macri signed Decree 133/ 2015 eliminating the 15% ET on beef, 15% ET on wheat, and 20% ET on corn. It also reduced 5 percentage points the ET on soybeans, bringing it to 30% and 27% for soybean-processed products. By the first trimester of 2016, sales had skyrocketed: beef was up 18% compared to the same period in 2015, wheat 69%, corn 63% and soybeans 37%. The five-point reduction on soybeans imply a direct reduction of $754 million[31] in the Federal Solidarity Fund (FFS), created by decree in 2009 to give the provinces 30% of soybean taxes. Despite the 5% reduction in the rate, the FFS rose 160% in the first trimester of 2016.

Infrastructure

Undifferentiated commodities such as soybeans have virtually no retail demand and so are overwhelmingly sold as inputs for processing. By strengthening command over other links in the production chain, trading companies have created a densely integrated network. Integration has occurred "upstream" with seeding companies and "downstream" with producers, furthering control over supply management and risk allocation. As a result of this mode of production, an equally flexible system of transport is being created. The geoeconomic pull toward a single, integrated "soybean republic" regional productive unit is noticeable in infrastructure. The main centers of gravity in soybean trade have a correspondence with the most efficient logistical points along the chain. The most profitable locations organize the spatial distribution of production and prescribe infrastructural improvements,[32] but geoeconomic dynamics have also impacted political centers of gravity. In Paraguay, where the locus of power has traditionally been local, the soybean boom has provided the resource base to further consolidate existing disparities. MNCs have benefited from this political economy arrangement and have taken advantage of weak institutionalization to further their interests. In Brazil, infrastructure developments have consolidated local engines of growth in the inner parts of the country. This has impacted political alignments, empowering local political figures that are not from the traditional Brazilian political alliance structure such as Governor Blairo Maggi. Finally, Argentina has also experienced "local empowerment." The soybean boom has generated infrastructural spillovers in lowering the costs of water freights, advancing what is the most modern and efficient vegetable oil processing complex and raising the strategic and economic influence of ports.

Argentina has its main soybean production zone in a radius of 300 km of the Santa Fe province ports. This setup gives the country a logistical advantage that has been leveraged heavily by traders and processors, building processing and storage facilities in the ports they control.[33] Nevertheless, in the face of mounting production and lack of the appropriate dredging, the port of Nueva Palmira in Uruguay is becoming an alternative loading station.[34] As soybean exports become more attractive, the radius has expanded (see map in Annex 2) with the associated cost increases. For example, in May 2016, Salta province PROGRANO producer association estimated that soybeans transported from Las Lajitas,

1,200 km from Rosario port, increased prices 25%. Another northern locality that has experienced high growth in soybean production—Las Maravillas—close to the Bolivian border at 1,500 km from port made the soybean cargo increase 26.6%. In August 2015, the shipping cost from Salta to the port of Rosario (1,191 km) was $72/tn. From Rosario to Shanghai, China (19,738.8 km) it was $41/tn.

In Paraguay, as soybean production rose, so did the improvement of the waterways and its facilities: new loading locations along the Paraná River were built and 2,400 km of water courses running to the port of Santos have been made navigable by the construction of three locks on the Tiête-Paraná waterway.[35] Shipments are half the cost of land freights (trucks) to Brazil. In addition, "multimodal" arrangements allow companies to make the time/cost cargo equation more efficient: through Itaipú, soybeans go up the Paraná River to Presidente Epitacio and from there are transported by train to the port of Santos. Waterways constitute the main avenue for transport of Paraguayan beans, spurring the development of a naval (barge) industry and promoting investments in port development. These improvements are badly needed, since the country has around 3,100 km of navigable watercourses. Infrastructure updates at the main port of Villeta, south of Asunción, cannot keep up with the continuous rise in soybean volumes. Navigational difficulties, lack of shipping space, and high freight charges (28.7% according to the 2014/2015 campaign averages) have complicated river transport to reach the Rio de la Plata basin. Moreover, unusually low levels in the Paraguay river led to fuel shortages and irregular services. Soybean-related infrastructure development in Paraguay does not signal an intention to foster broad-based economic development via spillovers into neighboring populations. Rather, they evince the result of a distributional struggle in which Paraguayan state institutions are caving to production needs of multinational corporations and foreign—Brazilian—producers. Since the inauguration of President Cartes in April 2013, the agricultural sector has made extensive use of the opportunities opened up by the new law on public/private alliances (APP). This has allowed an expansion and improvement of public infrastructure, like the project to dredge the Rio Paraguay. By widening and deepening the waterway, the aim is to make it more accessible and navigable throughout the year.[36] As a landlocked country, it is critical to Paraguay to make use of the Paraguay-Paraná river system as its major means of freight transport. With 84% of the Paraguay soybean exports transported by river, increased capacity of the

Paraguay river would allow more regular-paced shipments and reduce shipping costs while improving export potential for the country. Nowhere is infrastructure more critical than in Brazil.[37] Brazilian soybean production faces major infrastructural bottlenecks. The state of Mato Grosso is the center of the Brazilian soybean production. Former governor Blairo Maggi (2003–2010) was the single, most important soybean producer in the country. Despite the support, the agribusiness has enjoyed during his successors Silval da Cunha Barbosa (2010–2015) and José Pedro Gonçalves Taques (2015–2019), and Maggi was iconic of a flexible articulation between the political and economic levels in Brazil. The model of *"shared leadership"* between governments and businessmen in the agricultural sector promoted public/private partnerships to fund infrastructure projects the state. And the political debate has been structured thus in Brazil: many dismiss efforts as a covert attempt to advance his sectoral/family business and others point out to the fact that living standards have been raised throughout the state.[38] Because Maggi understood development as *"a shared responsibility,"* regulation is *"the product of consultations and cooperation between stakeholders and civil servants, in a continued dialogue to support strategic planning, guide action and facilitate project funding and execution."* The governor planned the *"verticalization of the economy: we do not have to export vegetable protein but animal protein. Our grains feed foreign animals; we should export the animal product. If we are to export soybeans, we should export them fried, cooked, packaged . . . with high added value".*[39]

Due to its size and distance between production and ports, Brazil would benefit the most from a transport cost reduction. By late 2015, transport as a percentage of total share was just under 10% for Campo Mourao (PR) and Río Verde (GO), in the 10–20% for Chapadão do Sul (MS), Sao Luis Gonzaga (RS), Balsas (MA), and Unaí (MG), 20–30% range for PDL (MT) and almost 40% for "agribusiness world capital" Sorriso (MT). Costa et al. estimated a 20% reduction would increase Brazilian soybean supply by 0.87–2.99% and lower the export price between 14.48% and 29.44%. In Brazil, in 2015, 63% of total agricultural cargo was moved by truck, along the 194.731 km of paved roads the country has. Only 26% is moved by rail and just 11% by barge. The big infrastructure projects in the region are guided by the soybean's economic pull. A known example is the paving of BR-163 road (Cuiabá-Santarém), in which Cargill is the principal actor interested. After modernizing the Santarém port, the company only needs an infrastructural improvement to this highway in order to efficiently get the soybeans from production sites to vessels with EU destination. The project has a total length of 976 km, comprising BR-163/MT/PA from junction with the MT-220 to junction with BR-230 (A) and BR-230/PA from junction with BR-163 (B) (Green Field) to Miritituba. According to the Logistics Investments Program (PIL), of the Ministry of Planning, BR-163 was in the public hearings stage by February 2016. More recently, BR-158/MT[40] is also emerging as an outlet for the Mato Grosso soybean production to be driven northeast. Within Pará, BR-158 is already paved, but in Mato Grosso, finishing its pavement is of critical importance to the government. Since resources are managed at the local level in Brazil, this situation has empowered governors and municipalities over the federal state, generating an alliance with multinational trading companies. As an example, in the state of Mato Grosso, the five main exporting companies are agribusiness companies. Together they are responsible for more than 60% of the state's exports: Bunge Alimentos (23.3%), Cargill Agrícola (12.9%), ADM do Brasil (11.4%), Amaggi Exportação e Importação (8.4%), and Louis Dreyfus Commodities Brasil (6.2%). An indicator of how the soybean boom has empowered private actors and how Brazil's state institutions have responded is the presence of public/private partnerships at the federal and state levels. Governmental decisions to allocate resources to a specific sector reveal the success of the political economy articulation of the agroindustrial chain to political power in Brazil. The new phase of the PIL program provides investments of $57bn: $25bn for railways (44%), $19bn for highways (33%), $11bn for ports (19%), and $2bn for airports (3.5%).

NOTES

1. After the 1973 "oil shock"—when Organization of the Petroleum Exporting Countries quadrupled the prices of oil—surpluses were deposited in international commercial banks, raising global liquidity. Oil-importing Latin American nations borrowed capital to finance the external deficits associated with oil inflation. The petrodollar recycling (peddling?) program allowed huge indebtedness. Loans also were different in composition: from long-term official loans with low interest rates to short-term commercial loans with variable high interest rates. Domestically, this debt financed growing trade imbalances. At the same time, the real exchange rate was kept strong to counter inflationary pressures. Continuous appreciation worsened the current account. The trigger came with a monetary contraction in

the USA. The (Chairman of the Federal Reserve Paul) "Volcker shock" raised interest on all US loans to 21%, bringing a sharp rise in world interest rates and a sustained appreciation of the dollar, which in Latin America manifested as recession, domino debt defaults, and capital flight.

2. In 1995, the top 10 soybean Brazilian processing firms were: Ceval, Sadia, Sanbra (which would later be Santista Alimentos S.A.), Cargill, Incobrasa, Unilever, Bianchini, Olvepar, Coimbra, and Coamo. By 1997, Bunge owned Santista Alimentos S.A. and had purchased Incobrasa and Ceval. Bunge then consolidated all processing operations under Ceval and all retail operations under Santista. At the same time, ADM entered the Brazilian market with the purchase of Sadia, and Glencore Grain Holding with facilities in Paraguay and Brazil. (Daryll E. Ray, *MidAmerica Farmer Grower*, Vol. 17, No. 47, November 24, 2000)

3. *Vital to the World*, John Rice (EVP, Commercial and Production) and Dwight Grimestad (VP, Investor Relations), Deutsche Bank Global Consumer Conference, June 10, 2009.

4. Type of vertical integration in which a buyer of raw materials acquires its suppliers, or sets up its own facilities to ensure a more reliable, cost-effective supply of inputs.

5. Bunge 2008 Annual Report, p. 7.

6. *How the Global Oilseed and Grain Trade Works*, prepared by HighQuest and Soyatech for the US Soybean Export Council and the United Soybean Board, November 2008, p. 13.

7. Officials made clear their unease with increasing agribusiness consolidation, with just a handful of firms controlling the lion's share of beef production, biotech seeds, and poultry growing.

8. The Associated Press reported Vilsack said: "*This is not just about farmers and ranchers. It's really about the survival of rural America. We've seen a significant decline in the number of farmers and ranchers and that translates into a significant decline in the number of people living in rural America.*" Sen. Chuck Grassley (R-Iowa) argued: "*Bigger isn't per se bad, but it can lead to predatory business practices and behaviors and that's what we've got to be concerned about.*" Under the banner "*Bust up big Ag,*" farmers complained that the lack of competition among biotech seed makers has led to a jump in seed prices, even as crop prices stagnate. Producers complained about the power corporations like Monsanto Co., Archer Daniels Midland Co., and Tyson Foods Inc. have over food production, while industry groups warned new laws or antitrust lawsuits could punish companies in the midst of a recession and stifle innovation and investment. Vice-president of industry affairs of Monsanto, Jim Tobin, said that the company's patented Roundup Ready gene has become the industry standard for the simple reason that farmers want it and that they can choose from competing

engineered traits offered by Monsanto's rivals. *"There's a lot of choice today, and there's going to be a lot more choice in the future, as the Roundup Ready gene patent expires in 2015."* "U.S. regulators examine competition in agriculture", AP, March 12, 2010.

9. Grain crushers carry out the separation of the two main components in the soybean: oil and protein (meal). Crushers or processors separate the oil from the flour. The crushing process is fairly uniform throughout the industry: soybeans are mechanically pressed to extract the oil. The residue is a firm hydraulically pressed cake (referred to as "soybean cake"), which can be later ground to form a loose meal. Crude soybean oil is then degummed (to remove the lecithin) and usually refined, bleached, partially hydrogenated, and deodorized to make a variety of popular product. For a complete list of soybean uses, see Annexes 2 and 3.

10. Heffernan, William; *Consolidation in the food and agriculture system*, Report to the National Farmers Union, Department of Rural Sociology, University of Missouri. Feb. 1999.

11. Bolivia, Chile, Colombia, Costa Rica, the Dominican Republic, Guatemala, Honduras, Paraguay, Peru, and Venezuela.

12. See N. 91.

13. An alternative to direct ownership is "time-chartering," whereby a trading company hires a ship for a period of time—generally six months to a year—and has extensive use of the vessel to move its product.

14. Total cost was estimated at $230 M. IFC would provide $40 M and the IDB would provide a $45 M syndicated loan and $70 million in equity and subordinated shareholder loans. Source: IFC Project Number 26.959, "Timbues Soybean Crushing Plant."

15. Los Grobo has been consistently pushed away from Argentina and forced to diversify into Paraguay, Brazil, and Uruguay. The Group estimates it will grow 30% and 20% in these last two coiuns, expecting a 0% growth for Argentina. "With this tax structure, the state ends up with 70% of our production" complained Gustavo Grobocopatel (*Fortuna* magazine, N. 330,09/25/09).

16. *Con la soja al cuello*, directed by Fabián Arocena, Redes—Amigos de la Tierra & Interludio, May 2009.

17. Brazil has a total of 250 multimodal terminals. The USA around 2500.

18. Known in Brazil as the "Soybean baron" (*Barão da Soja*), the governor was the largest individual soybean producer in the world.

19. Export duties are especially important in the case of soybeans from the BAP, which is mainly an export product with modest domestic consumption. As feedstock, soybean has historically been negligible in the Argentine, Paraguayan, and Brazilian beef industries since cattle was grazed in open ranges. Nevertheless, as the feedlot industry develops, soybeans may experience a more domestic-driven demand.

20. Beyond the obvious risks to any merchandise transaction (destruction, theft, disappearance, insolvency), crops have the additional risk of product spoilage or loss of condition due to storage and climatic conditions.
21. These features include: a) size, amount or quantity of the traded commodity, b) trading months and hours for delivery, c) product specifications (quality requirements), d) currency specifications, and e) minimum price fluctuation and maximum daily price limits.
22. As these pages are being written, MAtBA and ROFEX are conducting a preliminary feasibility study for the integration of both exchanges.
23. Created in 2008, BM&FBOVESPA S.A. is the union between the Brazilian Mercantile & Futures Exchange (BM&F) and the São Paulo Stock Exchange (Bovespa).
24. Many Paraguayan producers seemed to be rather satisfied with the "security" offered to them by having a steady buyer. However, some of them expressed concern at the fact that there is no way of making sales anonymous.
25. Storage bags for grains and forage. Silo bags can preserve up to 220 tons of crop for 12 months, regardless of weather. With a very low cost per ton of stored grain and efficient handling (turnaround time), silo bags were quickly adopted. A low initial investment is needed vis-à-vis metal or concrete silos, and there is no third party involvement. Moreover, grain stored in this fashion avoids paying taxes, since the BAP's tributary codes do not contemplate it.
26. Brazilian tax rules were compiled by lawyer Vinicios Leoncio and resulted in a work that weighs 7.2 tn and has 43,216 pages, each with 2.2 m high by 1.4 m wide.
27. Declarations to local news agency *Estado*, February 15, 2016.
28. "V Congreso de Economía Provincial", *Bolsa de Comercio*, Rosario, Argentina, September 24, 2009.
29. Sergio Bernesztein, private conversation, June 26, 2009.
30. "Lousteau reveló detalles de su paso por el Gobierno", *La Nación*, July 28, 2009.
31. Of these $754 M, $226 M are losses by provincial governments and $528 M by the national one. Partly is compensated by the recovery in gross income tax (IIBB), value-added tax (IVA), and earnings tax, so the net loss is not $754 M but actually $488 M, distributed in the same 30% to 70% proportions.
32. An example of this overriding force is Cargill's $20 M port in Santarém (state of Pará) that reduces transport costs to one-third. Because of environmental damage, the Brazilian Federal Tribunal ordered Cargill to close the port and pay compensation. Cargill filed an appeal with the Supreme Court and continued to operate. In March 2007, the Supreme Court ruled against

Cargill, and federal police shut down the port. Cargill ignored decision and resumed operations three weeks later. Due to the negative impact it has had on environment and human life, Franciscan Catholic priest Edilberto Sena calls it "the perverse scandal of the Lake Ness Tapajós river monster."

33. Argentine storage/production ratio capacity in 2015 was 40%, compared to 15% from Brazil and 60% of the USA.

34. Because Bolivia has no direct access to ocean ports, 49% of its soybean production leaves from Nueva Palmira, thanks to an operation concession first obtained by Siles Zuazo in the 1980s and later renewed in 1993 by presidents Paz Zamora and Lacalle.

35. This waterway is used by Brazilian giant Caramuru Alimentos—through a joint venture with company Torque—to transport its soymeal and lecithin to the port of Santos by tugboats and barges (with a 6000 tn capacity) to terminals in the city of Pederneiras or Anhembi (São Paulo). From there, a train to the port of Santos transports them. According to company Vice-president César Borges de Souza, the hydro-railway combination reduces logistic costs vis-à-vis the hydro-highway alternative of between 8% and 10%.

36. Usually, the Paraguay is navigable up through Asunción by vessels of up to 11 m draught. When in 2012 dry conditions lowered the river's depth, importers and exporters lost about US$250 million in potential revenues to navigation impossibilities.

37. Five ports concentrated 82% of Brazilian soybean exports in 2014: Santos 51%, Paranaguá 11%, Vitória 10%, Manaus 6% and Santarém 4%. All of them face severe constraints to being integrated in a wider transport system with road and railway access.

38. The Brazilian office of the United Nations Development Program reported Mato Grosso as the 11th state—out of 27—in the Human Development Index (HDI). The HDI increased from 0.601 in 2000 to 0.725 in 2010, a growth rate of 20.63%. The gap of human development—the distance between the HDI of the state and the maximum index value of 1—was reduced 68.92% in the same period.

39. Author's interview.

40. BR-158 crosses Brazil from north to south. It starts in Altamira, Pará, and ends in Santana do Livramento, Rio Grande do Sul, near the Uruguayan border. Its 3,864 km also cross the states of Mato Grosso, Goiás, Mato Grosso do Sul, São Paulo, Paraná, and Santa Catarina. In Mato Grosso, half of its 800 km are still dirt road.

CHAPTER 4

Coordination: Brazil

Abstract The Brazilian case is one in which local governance is much stronger, which has allowed to effectively integrate state institutions with the resource/sector (*coordination*). The main issues determining political economy coalitions in the country are the Amazon and land tenure during the rule of Fernando Henrique Cardoso, Lula Da Silva and Dilma Rousseff. The Brazilian model of governance of the soybean complex has created institutional adaptation that incorporates agricultural interests into the decision-making process in a mutually reinforcing relation. This is seen through the rise of a new axis of agricultural governors and the *bancada ruralista* (agricultural congressional bloc).

Keywords Agriculture · Agribusiness · GM seeds · Biotechnology · International political economy · Latin America · Soybeans · Brazil · Amazon · Lula · Dilma · PT

Soybeans brought an economic revolution to the Brazilian heartland. To what extent and in what ways has the productive structure brought about social and political change? The social impact is observable in the land access issue. The modernization of Brazilian agriculture coincided with the soybean takeoff in the 1960s. The national project sought to move Brazil away from the agrarian standard into a modern, industrialized mass society. Industrialization and urbanization demanded overcoming food

© The Author(s) 2017   83
M. Turzi, *The Political Economy of Agricultural Booms*,
DOI 10.1007/978-3-319-45946-2_4

supply and foreign reserve restrictions: modernizing agriculture served the elite's developmental strategy to boost food production and add new sources of dynamism to economic growth. As such, increases in productivity were essential for continued expansion of production. Modernization implied developing and adapting green-revolution technologies, and the government purposefully geared them toward large-scale agricultural operations that had important roles for mechanization and chemical inputs. This was in synchrony with domestic political objectives. President Juscelino Kubitschek's (1956–1961) modernizing imperative pushed the boundaries of agricultural production to redefine the structure and dynamics of the sector, including an export sector that relied heavily on coffee, cotton, and sugar. Since neither population, industrial, or urban sectors impose a large demand on the agricultural sector, it was possible to articulate such a political economy configuration, something impossible in Argentina during Perón.[1] Nevertheless, import substitution industrialization strategies posed a challenge to this model, since they required that the agricultural sector generates most of the economy's foreign exchange and guarantees domestic food supply. However, the transfer mechanism was via taxes on the foreign exchange earned by coffee exports. Through an overvalued exchange rate, depressed domestic agricultural prices redistributed gains in favor of the urban-industrial sector. This growth strategy, which was carried on by successive military regimes in late 1960s, required a fast expansion of exportable agricultural commodities. Carefully chosen agricultural chains[2] were chosen as technology-intensive agribusiness clusters. The government provided strong incentives for the creation and expansion of processing industries and agrochemical input industries.

The paradigm of Brazilian agricultural modernization was to induce transformation of the *latifundia*. This strategy was, politically, preferable to agrarian reform (Wilkinson 1997). Indeed, agriculture was part of a larger national economic development strategy that dates from the *Varguista* period in the early 1930s and whose implementation took six or seven decades (Barros 2009). Agricultural modernization also had a national security dimension: territorial occupation based on infrastructure expansion would ensure the protection of the Amazonian territory[3]: *Integrar para não entregar* (incorporate to avoid surrender). This expansion illustrates a hallmark of the Brazilian case: the state stimulated the interlocking of industry and agriculture. Strong, competitive agribusiness—as one of the economy's leading sectors—was firmly and continuously championed by public policy, for a number of reasons. The economic rationale was that agriculture was to serve as the primary source of generation of resources to be transferred for industrial development. A second, political objective was to populate the vast Brazilian territory, to secure borders and territorial integrity, especially in the Amazon. Finally, the national security reasoning, omnipresent in the military governments of the 1960s, was that an agribusiness organized agricultural sector would serve as a bulwark against radical land programs. As a result, the integration of productive chains into an agrifood system blurred the traditional industry/agriculture sector divide. Some authors (Muller 1989; Toledo 2005) claim it is no longer possible to identify an agricultural sector as an independent object of analysis. It is more accurate to speak of agroindustrial product-specific "complexes," each with its own combination of agricultural and industrial dynamics.

The main fault lines shaping the political economy configuration of the Brazilian soybean chain are environmental conservation and land tenure. The first one relates to the Amazon and has divided opinions along economy or ecology. The second one is embedded in the country's social structure and reveals the inclusion/exclusion dimension inherent to all processes of economic development.

The Amazon: Political Economy in Brazil's Far West

Total agricultural land in Brazil is 32.9 %, while forests areas amount to 61.9 %. And 60 % of the Amazonian rainforest is in Brazil. Deforestation has been steadily declining since 2004, according to estimates provided by the Brazilian National Institute of Space Research (INPE) and the UN FAO. The majority of the deforestation has taken place in the states of Mato Grosso, Pará, and Rondônia. Historical roots have important reverberations in the present, framing the issue of soybean production in the Amazon. Producers are themselves the result of internal migration motivated by the search for profit and a "new horizon." State encouragement of this process resulted in a "frontier mentality," much like the settlers of the American Far West[4] in the nineteenth century. Mato Grosso's own former governor Maggi—Minister of Agriculture in the Temer government of 2016—relocated to the state in the 1970s. At that time, in order to be eligible for loans to buy tractors and seeds, farmers had to clear land. In the field, the feeling of pride in all the producers is remarkable. There is a noteworthy sense of authority and dignity derived from having conquered nature and achieving progress for their families and their countrymen. Maggi felt it a personal accomplishment to have turned the state from a malaria-infested zone into a world-class agricultural powerhouse.

Contrary to the image of nature as something to be protected, government officials of soybean-producing states frame their discourse on the forest as backward and brutish, an entity to battle against. In a 2009 interview, then Mato Grosso's secretary of agriculture, Neldo Egon Weirich said, it was *"revolting and sad when the world said that deforestation was bad; we were told to come here and that we had to tear it down."* Even with Federal laws stipulating 80 % of tracts in the upper Amazon and 35 % in the *cerrado* regions must remain forested, Brazil has problems because low enforcement creates an incentive to cheat. Brazilian Amazon deforestation rates escalated in the early 2000s. After reaching a high at over 27,000 km2 in 2004, it then decreased sharply to about 4,500 km2 in 2012, according to 2015 data from National Institute for Spatial Research (INPE). Leaving the rainforest intact is not an option for producers. As president of the Mato Grosso Farm Bureau (FAMATO), Rui Prado explained it: *"To condemn a whole region to backwardness is not fair. No production means no energy, no roads, no railroads, and no ports. Without the state we would have no taxes, but also no infrastructure, or governance."* On the same occasion, former Governor Maggi had declared *"the jungle is not better; the jungle is synonym of backwardness."* An instinctive suspicion reigns on the environmentalist/conservationist movement, coupled with the fact that they are "outsiders" (whether from the city or from other countries), to form the notion that they do not understand the issues correctly and cannot appreciate the hardships producers face. Former Director of National Transport Infrastructure Department and former Mato Grosso state infrastructure Secretary Luiz Antonio Pagot enthusiastically supported the view that environmentalism is an international conspiracy against Brazil's sovereignty and growth. Notwithstanding, since 2005, the state of Mato Grosso alone has reduced its deforestation more than 80 %, and has been responsible for more than half of Brazil's 70 % reduction of deforestation in the Amazon region.

"A farmer in red cannot take care of the green" said Ricardo Arioli Silva in 2008, the then President of the Mato Grosso Association of Soybean Producers (APROSOJA). The phrase crystallizes how the debate about the impact of soybean production has been framed in Brazil. Tree huggers vs. golden chainsaw awardees, conservationists vs. profit seekers, green pacifists vs. strong *fazendeiros*. Deforestation and its effects are beyond the

scope of this book. Although it constitutes one of the hidden costs of agricultural production, cattle ranching rather than soybean are the immediate driver behind deforestation. Indeed, Environment Ministry (MMA) figures place that industry as accounting for 78 % of new forest clearing. The second place is taken by illegal logging. It is undeniable that the industry has an enormous level of responsibility.[5] There are not only connections between the industries[6] but also an inter-temporal dimension to this problem. The frontier advancement cycle is: clearing (cutting down the forest), logging (selling the native woods), pasture creation for cattle, and use of the grasslands for crop production. Both ranchers and loggers are also condemned for using violent tactics against indigenous populations such as intimidation and even murder.[7] Hired gunman or *jagunços* in Brazil have killed 1,934 rural workers and activists in Brazil since 1985 according to the Pastoral Land Commission (PLC) of the Catholic Church. Charges were brought against them in less than 10 % of those cases. Convictions are non-existent or quickly overturned. In 2015 there were 50 activists killed, 45 in the Amazon alone, making it the most violent year since 2004, according to the CPT annual report. Violence is legitimized as a normal part of politics. It has become informally accepted. Former deputy editor of National Geographic Brazil, Felipe Milanez declared in February 2016: "*I've never seen, working for the past 10 years in the Amazon, a situation so bad.*"

Although agricultural policies and regulations are set at the Federal level, soybean-producing states have followed a proactive policy stance. While states cannot legislate on agricultural issues, they can be harder or more conservative in establishing higher standards than those set by the Federal government. Although there is great heterogeneity within and among regions, the functioning of the whole seems to be an indication of how Brazilian federalism works, with each state managing the responsibilities of decentralization after the 1988 Constitution differently, both in geographic and in sectoral terms. For soybean-producing states such as Mato Grosso, state capacity depended greatly on the sources of revenue generated by agribusiness. Thus, enhancing this economic engine became an economic policy necessity and a political imperative. This dynamic is a positive externality of the Brazilian Federal system that is absent in Argentina and Paraguay. In these two countries, the distributional struggle is intensified by the absence of such revenue-sharing institutional arrangements. In Argentina's case, it heightened social and political conflict by an unresolved parity of forces, while in Paraguay, it led to state capture by overwhelming the agricultural governance structure.

Has soybean production increased or decreased governance? As John Carter of Aliança da Terra expressed it:

> Soybeans bring governance and logistics (via infrastructure development), and less fires. Asphalt brings competent people and law abiding ones, contrary to what environmentalists argue. The Amazon is a frontier, and it is a nightmare. Brazil has a sovereign right to develop this frontier, and frontier is synonym of lawlessness, and if you obey the law then you are being silly. There is no villain; this is a process only solvable by economics. Command and control does not work; enforcement issues are too big. Only 5% of the Amazon land has official titling, and [environmental] demands are on this 5%. They are so impossible that they lead to a de facto civil disobedience. Governance is the key and is sustainable only through economic incentives. The forest is a disincentive to guarantee property rights, without it, it is easier to enforce them, plus the land becomes more valuable.[8]

While governor of Mato Grosso, Maggi became a "free-market environmentalist." He has an economic approach to deforestation and advocates a game-changing strategy: paying farmers to keeping the forest standing. The financial reward lowers the incentives for deforestation, and—should a carbon credits market be developed—could become a self-enforcing effort against global warming. So far, it has been the most viable alternative anyone (including environmentalists and international agencies) has come up with. There has been concrete delivery on the part of the Mato Grosso government.[9] There is an economic rationale for deforestation: forests do not yield profit,[10] and there are no incentives for keeping them standing. Only initiatives that acknowledged the overriding incentives created by production vis-à-vis environmental benefits will in the future be effective. This perspective may explain why the soybean moratorium, which increased prices via supply reduction and demand for certified beans, has been effective. Economic incentives are self-enforcing and transform producers into stakeholders of a system that penalizes environmental cheating.

What underlying aspects, then, does the struggle over the Amazon illuminate for our political economy analysis? A straightforward example is provided by the economy/ecology debate, epitomized in the clash between former environment minister Marina Silva and governor Maggi, since their rift exemplifies the main dimensions of the issue. Silva is a renowned environmentalist, a former rubber tapper to be the first to be elected to Congress. She was portrayed as heir to Chico Mendes' struggle that fought her way through adversity and poverty with honesty and hard work. When she decided to take on agribusiness, her public image started to be molded exactly into how agribusiness representatives view the ecological movement.[11] Silva was presented as frail, naïve, and incapable to understand—let alone solve—the strategic importance of the sector to propel Brazil into development and global standing. Maggi, on the contrary, not only epitomizes agribusiness interests, but also a way to do business. He is a pragmatic agrarian capitalist and does not care much for restrictions or "interventions" to the free hand of market, even at the expense of a hefty environmental cost. Almost 20 % of Brazil's Amazon and 45 % of the *Cerrado*—the Brazilian savannah—are deforested. The main tributaries of the Amazon river from the eastern border of Peru and the Andes have around 150 proposed dam. In the southern banks, big rivers—Tocantins, Xingu, Madeira, Teles Pires, and Tapajos—have been or are being dammed. In 2015, there were 40,000 fires in indigenous lands just in Maranhão state. The Xingu Indigenous Park in the state of Mato Grosso, surrounded by soybean plantations, is burning every year.

In Brazil, even when a consummate environmentalist like Marina Silva occupied the Ministry of the Environment, a government-agribusiness coalition quickly neutralized her most bold policy proposals. The rural sector's economic interests have successfully interwoven with a flexible mode of political representation. The institutional expression of this political economy is the *bancada ruralista* (BR) or congressional rural bloc. BR membership by Feb 2016 was 207, 40 % of the House. However, since many congressmen have family ties to agricultural production and the BR only gets together in issues pertaining to the agricultural sector, the vote count can be higher. Indeed, other caucus such as business (208), contractors and construction (226), or relatives (238) have many points of interest with the BR. Costa (2015) estimated 351 parliamentarians have a total of 863,646 hectares of land in their names, in the name of their families, or through companies which are owners or shareholders. The bloc started to gain visibility in the political scene after the 1987–1988 Constitutional Assembly, as a continuation of the contentious Ruralist Democratic Union[12] (UDR). At first, the BR was formed by UDR people, but in the 1990s, the Confederation of Agriculture and Livestock (CNA) and the Brazilian Rural Society (SRB) had representatives elected to Congress, fragmenting the political representation of agricultural interests. Indeed, current CNA president is also Senator Kátia Regina de Abreu (TO), whose campaign was funded with CNA money.[13] The BR adopted the two-pronged strategy of funding and favors: funding those representatives willing to defend "the interests of modern agriculture and private property" and establishing reciprocity alliances in the chambers. The agrarian elite understood the importance of organizing as a pressure group for lobbying, but did it in a novel way. Instead of exerting influence on parliamentary representatives, they became representatives themselves, creating an ad hoc coalition of corporatist representation. Members come from every party, from the more leftist PTB (Brazilian Labor Party), the PMDB (Brazilian Democratic Movement Party), PSDB (Brazilian Social Democracy Party), the DEM (Democrats), and the bloc has received the support of the ruling PT (Worker's Party).[14] Gaining political space, to the point of becoming the largest interest group in the Brazilian state[15] (Vigna 2007), was possible through the extraordinary flexibility of the Brazilian political system.[16] As deputy Valdir Colatto (PMDB/SC) stated: "*When we work for the agricultural issues, there are no parties. The party is the agricultural because we defend agricultural productive interests. Although we are from different parties, we all work for the agricultural sector. For the legislator who was elected directly in agricultural zones, the party has no weight whatsoever.*"[17]

Tactics combine alliances with sectoral (federal and local) organizations and industry leaders, lobby, and reciprocity through an extensive and diversified influence network and the control of key posts in the decision-making process.[18] The BR votes as a bloc only when its interests are at stake, leaving its members freedom of action for the rest of congressional initiatives. Thus, the BR is at the same time a bloc of power in Congress and a political subject enmeshed in the Brazilian state structure. Consolidation of the BR as a power bloc rides on the flexibility of the Brazilian political system, where switching labels is costless. As explained by opposition Deputy Ronaldo Caiado (DEM/GO): "*When the issues are of interest to the agricultural sector, we have the support of the ruralista part of the ruling party. So the bancada ruralista stops any governmental attempt and can negotiate in favor of the agricultural sector.*" To which Colatto added: "*Ah, but we in the ruling party also defend the farmers, for we go beyond party issues and help the government resolve the sector's problems.*"[19] This flexibility would not be possible in the more rigid Argentine political system of party loyalty[20] and not necessary in the Paraguayan political system of limited competition.

More structurally, the BR also signals the ascent of a new landed elite from the Brazilian interior, globally connected and locally grounded.

The agribusiness sector has grown largely self-sufficiently from São Paulo. Consequently, it has developed a self-reliance from the economic and political clout the city has traditionally enjoyed over the rest of the country. In Argentina, a parallel is observed in a rural sector whose development hinges more on its intrinsic competitiveness rather than on its dependence from Buenos Aires. However, the more atomized Argentine agricultural structure has made representation less unified than in Brazil. The Paraguay case is similar and different at the same time. While local landowners are powerful, of, there is no central government or industrial center to oppose. Their support comes "from above", with the Brazilian government pressing its Paraguayan counterpart, and "from below", with the Paraguayan landlords represented by the Colorado majority in Congress and in key bureaucratic posts. The Brazilian *bancada ruralista* style of representation showcases a growing capability for linking the different regional forces with no need of "mediation" from the *paulista* center. Although it could be argued that this political position is riding on buoyant international prices of agricultural commodities, all the evidence points to the BR increasing rather than losing power. Furthermore, should one of the *ruralista* figures be launched into the national scene, it could have the potential to shift the traditional axes of power and influence represented by the PSDB. However, such a move toward restructuring political forces would imply a broader agenda on the part of the *turma da botina* (literally, boot group), the political faction and operatives of governor Maggi. So far, the focus of representation has been kept sectoral and circumscribed to strengthening its hold on Mato Grosso.

Land Struggles

A historical line connects the Amazonian issue with the question of land access and land tenure, as the ecological and social dimensions of a process of economic development and territorial expansion. Brazilian distribution of land and of rural income has been concentrated since colonial times, and soybeans have not altered this political economy pattern. As Alston et al. (1999) compellingly argue for the states of Pará and Paraná, land policies have led to the creation of a self-reinforcing circle of concentration and conflict. Brazil's spasms of intense yet incomplete land reform are the result of a set of policies that respond less to a coherent strategy of agricultural development and more to piecing together the multiple interests which federalism incorporates in the policymaking process.

Land reform as a concept has a normative, sociopolitical component (distribution to ensure equality of access) and an operative, economic driver (unlocking the latent potential of the poor rural households). As identified by Carter and Zegarra (2000), the land reform agenda can comprise any degree and combination of its four components:

(a) Legalization of land tenure (secure ownership through property rights)
(b) Transaction costs reduction in land markets to facilitate private contracting between small and large landowners (rentals)
(c) Market-assisted land redistribution through state subsidies to help the poor overcome capital scarcities that act as barriers of entry to land markets
(d) Competitiveness enhancement of small units of production through state assistance.

The Brazilian Agricultural Research Corporation (Embrapa) compared the number of agricultural properties in the 1985 and the 1996 census and found the number shrinking by 16.3 %. However, for soybean properties, the retraction was of 42 % (from 420,204 to 242,998). Data from the latest national agricultural census 2006 shows high levels of land concentration, both in ownership and production terms. Measured by number of farm units, 73.7 % of total ownership is found in establishments in the segment between 10 and 50 hectares. Yet, owners of establishments of more than 100 farm units represent only 1.6 % of the total. The Brazilian Institute of Geography and Statistics (IBGE) estimates soybean production occurs on more than 171,000 individual farms across the country. Distribution of these soybean lands in terms of land area per owner is more unequal than that of cattle ranching and staple crops—rice, manioc, and beans. However, it is less unequal than that of sugarcane production (Garrett and Rausch 2015; Lapola et al. 2013; Martinelli et al. 2010). The concentration of land in the hands of fewer people reinforces rural income inequality, as there are few off-farm employment opportunities. The FAO's Gini index of land concentration for Brazil is of 0.87, with a regional average of 0.83 (Argentina stands at 0.83 and Paraguay at 0.93). Furthermore, in unstable macroeconomic environments land is used not only as a productive asset but also as a hedge against inflation, as a capital investment to be liquidated in order to smooth consumption in volatile downturns, collateral for access to agricultural loans or as a low tax bracket investment. Brazilian agriculture has always been a tax shelter. A long history of macroeconomic instability led to land prices to increase because of its position as an asset protected from taxation, especially in periods of hyperinflation, when land ownership provided very good protection against inflation tax. In the 1990s, as inflation rates rose throughout the region, so did the hedge value of land. Reduced uncertainty relieves the speculative pressure on the price of land, decoupling it from the capitalized value of the income stream generated from agriculture. As a result, the Brazilian case exhibits an underdeveloped agricultural rental market.[21] Less than 3.3 % of Brazilian agricultural land was under lease or sharecropping contracts in the latest—2006—World Census of Agriculture, dated from 2006. As demonstrated by Macours (2004), insecure property rights reduce the level of activity on the land rental market by raising the fear of loss of property when not traded within a narrow local circle of confidence. The author shows how the tenancy market is matched along socioeconomic lines, limiting access to land for the rural poor. Furthermore, Barros (2009), based on data from IBGE, shows the real price of land decreased 50 % between 1989 and 1999, but then increased back 70 % by 2007. Steward (2007) found that where soybean production is profitable and land for expansion scarce, land prices rise very quickly. Furthermore, peasants can sell to avoid being enclosed by the big soybean *cultivares* that—in the process of expansion—break the social networks that have been broken by their neighbors' land sales (Steward 2007; Baletti 2014).

Quite the opposite happens in Argentina, where the balance of power is in favor of land tenants rather than owners. Despite law 13.246/1948[22] for regulating land rentals, rural rent prices were time and again frozen, in order to protect rural tenants from the abuses of landowners. The government sided with tenants against owners, the opposite situation of what has happened in Brazil. This scenario led to adaptive strategies on the part of the actors, and traditional rentals were progressively abandoned in favor of more informal arrangements. Furthermore, in order to capture gains from scale—particularly important in the case of the new soybean production model—producers achieved land exploitation without recurring to land property extension. Either producers' cooperatives or—as is more frequently the case for soybeans—financially backed sowing pools rent out adjacent productive units. As the case of *Los Grobo* demonstrates, this land rental flexibility is at the heart of the competitiveness of Argentine soybean production. A third variant closer to the Brazilian experience is the Paraguayan case, in which gains from scale are obtained by a classic pattern of ownership concentration that rides on the exclusion of increasing parts of society, namely landless peasants and indigenous populations. Because President Stroessner (1954–1989) distributed land in a discretionary (neopatrimonial) fashion, the development of a land rental market was virtually impeded. Low rental values ultimately bring down the price of land, which encouraged large-scale ownership and brought together concentration. The end result is that today the Paraguayan soybean model of production is more limited in its economic and distributional potential by a more rigid and regressive agricultural structure.

Land tenure in Brazil is also determined by the needs of the model of soybean production. Measured by area, establishments of over 1,000 hectares accounted for about 43.8 % of rural land in Brazil, while those of less than 10 accounted for only 1.8 %. This dimension is inherent, but not exclusive, to the soybean model of production. Previous large-scale productions in Brazilian agriculture—such as sugar and coffee—created similar patterns of landholding as soybeans.[23] Because of the incidence of the technological component explained in Chapter 3, agricultural production was brought into the knowledge-based economy (Drucker 1992). On the one hand, the ascendancy of the soybean technological package implied higher entry barriers and the consequent tendency to concentration. The number of producers is reducing and scale is needed to support local fiscal pressure. At the same time, it has spurred backward linkages to other sectors of the economy. This has transformed agricultural production into a technology-intensive, export-oriented activity, generating a demand for specialization that created new rural social strata: contractors operating machinery, professional service providers for seed adaptation, and consultants who design and manage agricultural production systems. In fact, modernization without redistribution has arguably been the political objective of expanding soybean production for the last three decades in Brazil. In the 1964 Land Act, the military government set the foundations for regulating land access and tenancy in order to assuage pressures for land redistribution. Peasant leagues, organized by rising grassroots Catholic priest *tercermundista* (third world) activism, were perceived as potentially conducive to agrarian rebellion. The government's answer was to upgrade large landholders by subsidizing soybeans. Rural credit availability for this type of production resulted in the absorption of small farmers by medium and large properties, concentrating land distribution even further. When the sector experienced economic liberalization in the 1980s, the 1988 Constitution required that land serve a social function.

The 1988 Constitution's "social function" clause asserts that land can be taken by the state if it did not meet any one of four distinct conditions: (1) "rational and adequate use" of that is done such as to ensure (2) the "preservation of the environment", (3) the "observance of provisions regulating labor relations, and (4) "exploitation that favors the well-being of owners and workers" (Art. 186).

With the instability of the Collor administration (1990–1992), the process was halted. In his first term (1995–1998), President Cardoso accelerated the rhythm of the settlements, redistributing between 7.5 M (National Institute for Colonization and Agrarian Reform—INCRA figures) and 12 M hectares (official Presidency figures), and much more spread across the Brazilian territory. In spite of this official policy, the period was characterized by conflicts and land invasions,[24] associated with the Landless Movement (MST), the largest social movement in Latin America. By "direct action" (land occupation), the MST resettled hundreds of peasants on active or fallow *latifundias*, gaining visibility and pushing agrarian reform onto the national agenda Thus, in his second administration (1999–2002), Cardoso refocused land reform. Cardoso "allowed" the MST to do a social, non-state, bottom-up, direct land reform. It was politically less costly to let the MST take care of the complex, but was controversial and difficult task of appropriating rural property from politically powerful landowners than for the government to carry out the task itself. However, MST's land occupations and protests embarrassed the Cardoso administration, making it vulnerable to criticism that the administration was not protecting "law and order." The President criticized the MST in public while simultaneously settling a significant number of families: 461,066 over eight years. During the Lula years (2003–2010), the land reform issue received the *coup de grâce* by the fracturing and weakening of the landless movement. MST was also an integral part of the social movements' coalition that brought Lula to the Planalto palace in 2002. Lula's subsistence federal food allowance to the extremely poor program (*Fome Zero*) effectively "crowded out" the leftist side of the political spectrum. Increased social expenditures reduced support for direct action tactics. Co-optation of labor leaders isolated the MST from its urban ties, fracturing the movement's representation spectrum and hindering their capacity to protest and strike. Although the PT refused all lobbying from agribusiness and other powerful sectors of Brazilian society to undermine or even criminalize the MST, it did not exert pressure over state governors, local oligarchs, or the judiciary in their defense of agribusiness. Lula sought to bring together its diverse constituencies to work out a compromise. He appointed Roberto Rodrigues—an agribusiness advocate—as Minister of Agriculture. The PT did to address land and wealth concentration in the hands of the rural elite because agribusiness promised increased revenue from large-scale export products. It also appeared key to diversify the energy matrix of the country, to gain a growing share of the biofuel market. Lula did not take the politically charged step of expropriation for the purpose of land reform. The plan to settle one million families was cut in half, although softened with provisions to legitimize the land claim of thousands of Brazil's most vulnerable rural workers, the *poseiros*.[25] The PT government sided with agribusiness as a model of production and governance, encouraging its growth with subsidies, tax breaks, tariff breaks, and other incentives. Carter (2015) notes that large-scale corporate estates received an average US$356,729 in subsidies, while the average family farmer US$9,079. Between 2003 and 2012, the CPT identified 63,417 cases of enslaved workers and 2,569 landowners accused of serious labor code violations. Yet, the only criterion used by the PT government for state expropriations was the first one of the four mentioned above. This implied a de facto *Expropriação Zero* policy: as land was productive no amount of ecological damage or labor exploitation would—in practice—justify expropriation. The soybean model of production has gone beyond mere agricultural modernization: it transformed and unleashed a revolution in Brazilian social relations that sustain the sectors' political economy. It has empowered the *sector agro-fundário* (landed agricultural sector) and the exporters and traders over the indigenous populations and the landless peasants. From 2003 to 2014 there were 390 Indians killed in Mato Grosso do Sul, mostly Kaiowa Guarani, fundamentally in conflict with ranchers and soybean plantations. Land redistribution stagnated during the Dilma Rousseff administration. In her first mandate (2011–2014), she completely ignored agrarian reform, to the point it was not even included in her signature campaign, *Brasil Sem Miséria*. During this period, the number of settled families—26,557— was less than under Cardoso, Lula, or even Collor (1991–1992), when the total amounted to 37,493.[26] Founder and coordinator of the MST, João Pedro Stédile said in 2012 the Dilma government "had abandoned agrarian reform" as it "could not even solve the social problem of 150,000 families encamped, some for more than five years, along Brazilian roads".[27]

The land issue in Brazil is channeled through the following actors and institutional arenas: landowners, landless peasants, the INCRA, and the courts (Alston et al. 1999). Landowners have defended their interests through the bancada ruralista (BR), their congressional arm. The Brazilian ruling class does not have hegemony—in the Gramscian sense—in the Federal government. Participates, but does not command. Thus, it has chosen the local and judiciary to secure their rule. Municipal and state authorities did de facto cohesively criminalized occupation, channeled through the judiciary. Courts are the arenas in which land conflicts are played out, where landowners have challenged the Federal governmental level, typically resisting INCRA. The process of land disappropriation begins with an act signed by the Brazilian president but always finishes with a judicial decision (Reydon 2000). The state needs to pay a "fair price," which after the 1993 amendment became the "market price." The decision must address the items to be compensated, the amount of the compensation, and the form of payment (public bonds vs. cash). On these grounds, landowners have systematically—and successfully—challenged expropriations, claiming the price offered was too low, that the land did fulfill social functions, that inspection was not conducted properly, or that INCRA never notified owners directly. Delaying tactics were effective: the longer the process, the more expensive for INCRA and thus the more likely the case would be dropped. INCRA estimated the final cost of an expropriation at five times the initial evaluation.[28] The consensus among landowners is that INCRA's actions are not conducive to achieving the social benefits that could be derived from land reform, for it relocates people around the country without generating provisions for sustaining them once there. Without a working relationship with local governments, basic service provision (water, electricity) is not possible, and seeking the most immediate resource to sustain him, the migrant turns to deforestation to generate income. For the landless, INCRA falls short on the other end. It has been accused of not standing up with enough authority and political will to the landed interests, of moving too slow and of not being capable of delivering effective occupation of new territories. Moreover, there are serious corruption accusations on INCRA's involvement in *grilagem* (land-grabbing)[29] and *laranjas* (oranges)[30] frauds. INCRA has diverted "resettlement funds" assigned with each parcel.[31] Lack of a unified land registration system and coordination between real estate notary offices and agrarian government agencies at the municipal, state, and federal levels create the loopholes that allow this practice to endure.[32] In one of its most outlandish twists, the Land Institute of the state of Pará announced that in four of its municipalities—Moju, Acará, Tomé-Açu, and São Félix do Xingu—registered surface exceeds actual physical area.[33]

Notes

1. In 1946, President Farrell—following Colonel Perón's recommendations—created the Argentinean Institute for the Promotion of Exchange (IAPI), which instituted a state monopoly to manage agricultural exports and the importation of strategic products. The aim was to centralize foreign trade and transfer resources between economic sectors.
2. Soy meal and oil, instant coffee, processed beef, poultry, orange juice, sugar, and alcohol complexes received subsidized credit, guaranteed prices, and tax exemptions and subsidies when exported.
3. See one of the *Marcha para o Oeste* (Westward March) Amazon development campaign governmental flyer in Annex 5.
4. This "Far West mentality" is part of the pioneer culture, and it is indicative of an absence of the rule of law. Celebrating gun-toting as a demonstration of strong leadership or effectiveness is not to be attributed to some racial or cultural Latin American trait. Sarah Palin could readily fit into this leadership style.
5. In 2006, the Brazilian Vegetable Association of Cereal Exporters (Anec), the Brazilian Vegetable Oil Industry Association (Abiove), and the Brazilian Grain Exporters Association (ANE), together with Greenpeace and WWF formed the Soy Working Group and decreed a moratorium on sourcing from newly deforested areas of the Amazon, an initiative to curb destruction of the rainforest. The industry pledged member companies would not trade soy originating from areas deforested after July 24, 2006. The two-year agreement was then extended until July 28, 2009, when it was renewed for a second time, remaining in effect until July 2010.
6. At Bun fields are too high. Bunge absorbs the region's whole soybean production, processing 500,000 annual tons of soybeans. The company argues the Uruçuí unit has generated 120 direct jobs and 10,000 indirect ones. What it fails to report is that it received soybeans from Fernando Ribas Taques, owner of the Carolina do Norte *fazenda* in Alto Paranaíba (MA) even after entering the *lista suja* in December 2006. To enforce the National Pact for the Eradication of Slave Labor, the Federal government has created a "dirty list" (lista suja) to publicize those commodity producers (individuals and companies) that resort to this heinous practice. The aim is to prevent them from selling or obtaining credit. The list is available online: http://www.reporterbrasil.org.br/lista-suja.
7. One of the most notorious incidents was the assassination of American-born Brazilian nun Dorothy Stang on February 12, 2005 in Anapu, state of Pará. Rancher Regivaldo Pereira Galvão was convicted of ordering the killing. Lawyer Americo Leal based his defense on challenging that Sister Stang was indeed a nun, portraying her instead as an agent of the US government sent to create destabilization in the Amazon in order to later "colonize" it.

8. *Agribusiness and Sustainability in Brazil Farming in Mato Grosso, The Border of the Amazon*, Woodrow Wilson International Center for Scholars, Washington DC, April 12, 2008.
9. See "Smarter farming key to saving Amazon rainforest", *The Associated Press*, February 7, 2010 and "In Brazil, paying farmers to let the trees stand", *The New York Times*, August 21, 2009.
10. Understanding that, in the absence of a financial reward, there is no incentive to keep the forest standing is one way to solve the issue of deforestation. Offering yearly cash payments to keep the rainforest is a rather simple and effective solution.
11. Incapable of integrating their interests and views in the national agenda, Silva lost her battle against agribusiness, left the MMA, her party (President Lula's PT) and switched to the Green Party, under which label she could be running for office in 2011. When Silva left the MMA, a deputy from the *bancada ruralista* told the author: "*chegava a dar pena daquela moça tão bem intencionada...*" (I pitied that little girl, she was so well intentioned...).
12. Founded in 1985, UDR's objectives during the assembly were to block agrarian reform initiatives and obstruct any kind of land redistribution, maintaining the power of landowners.
13. Documents proving this were published in "Tem boi na linha", *Veja*, No. 2066, June 25, 2008.
14. "Com aval do PT, ruralistas dominam comissão do código florestal", *Folha do Sao Paulo*, October 14, 2010
15. In the 2007–2011 Legislature, Vigna estimates the BR occupies 23 % of the Lower House.
16. There are also important background conditions. The emergence of biofuels as the key for a new Brazilian energy matrix and a cornerstone of the developmental strategy creates favorable structural conditions for these interests. Combined with the rapid expansion enabled by GMO crops, agribusiness power consolidation is the result of their weight in a political project and a growth strategy.
17. *El Parlamentario* magazine, December 19, 2008.
18. This is the case of senator Abelardo Lupion (DEM/GO), who voiced the interests of the BR at the Joint Parliamentary Investigatory Commission (CPMI) on Land Reform, where he proposed bills that typified collective land occupations as "abject crime" and "terrorist acts".
19. *Ibid*
20. When Vice-president Julio Cobos voted against the sliding tax in July 2008, he became a "hero" to the agricultural sector and a "traitor" to the government. In August 2009, senator Roxana Latorre was instrumental for the government to pass the agricultural emergency law. Her switch (she was

elected in June under a pro-*campo* platform) fractured senator's Reutemann (currently among the top contestants for the presidency in 2011) bloc. Such a switch was viewed as "treason" by Reutemann and as a maneuver to hurt his electoral chances.

21. In 1964, Brazil adopted the Land Statute, which imposed several binding and non-renounceable clauses that aim to benefit the sharecroppers and lessees.
22. Modified by law 22.298/1980 and not amended by late 2015.
23. The similarities between the landowners of the coffee plantations in the passing of the 1850 Land Act (which prohibited land acquisition through squatting and limited it to purchase) and the *bancada ruralista* in its struggle against the MST are striking.
24. One of the boldest movements came on March 2002, when nearly 600 members of the MST invaded the *fazenda* Córrego da Ponte, property of the Cardoso family in Buritis, Minas Gerais. The issue was quickly politicized in the wake of the presidential campaign.
25. Families that have rented (possessed) the same pieces of land from large landowners for many generations.
26. "Dilma assentou menos famílias que Lula e FHC; meta é 120 mil até 2018". Globo.com, March 30. http://g1.globo.com/politica/noticia/2015/03/dilma-assentou-menos-familias-que-lula-e-fhc-meta-e-120-mil-ate-2018.html
27. Interview, *ABCD MAIOR*, 12/01/2012.
28. *Livro Branco das Superindenizações, Ministério do Desenvolvimento Agrário*, available at http://www.incra.gov.br/portal/index.php?option=com_doc man&task=doc_details&gid=319&Itemid=273
29. *Grilagem* refers to an illegal acquisition of property and (public) land by means of forged documents. Originally, they were left in a cricket—*grilo*—packed box to give the paper an old appearance. Today, more sophisticated methods are used, like altered satellite images and GPS.
30. The *laranjas* entail the adjudication of land parcels to phantom owners, whose name is really a front for loggers who buy adjacent properties that they later deforest.
31. Former Federal public prosecutor Mario Lucio Avelar stated the national average fraud in land redistribution was 53 %. In the state of Mato Grosso, the number climbed to 80 %.
32. São Paulo University professor Ariovaldo Umbelino de Oliveira estimated the total area usurped by this practice to be of 171,605,152 hectares. When INCRA reviewed Amazonian land ownership records, it found that more than 62,000 claims appeared to be falsified.
33. http://www.iterpa.pa.gov.br:8000/ascom/index.php?option=com_con tent&view=article&id=13:instituicoes-publicas-do-para-pedem-cancela mento-de-titulos-falsos&catid=19:ascom-antigas&Itemid=16

CHAPTER 5

Colonization: Paraguay

Abstract Paraguay exhibits a dire pattern of dual economies, where indigenous and landless peasants are systematically marginalized. The soybean model of production is secured institutional capture from private economic interests. In this case, there is a coalition of colonizing interests: MNCs and Brazilian and *brasiguayo*—Brazilian settlers in Paraguay and their descendants—landowners, overwhelmingly supported by the Paraguayan landlords. Although the formal structure is that of a unitary state, the agricultural sector has achieved de facto decentralization by state capture. Taking advantage of power asymmetries and weak initial institutional conditions, there has been *colonization* by particular and foreign interests. These are the determining factors of the political economy coalitions during the Lugo and Cartes administration.

Keywords Agriculture · Agribusiness · International political economy · Latin America · Soybeans · Paraguay · Brasiguayos · Lugo · Cartes

Paraguayan author Augusto Roa Bastos described his country as *"an island surrounded by land."*[1] Indeed, Paraguay's isolation was the result of conscious policy decisions. Surrounded by hostile neighbors who repeatedly attempted to dominate the country in the early 1800s, the leadership judged self-sufficiency to be the path to self-government and sovereignty. From 1814 until 1840, Dr. José Gaspar Rodríguez de Francia

© The Author(s) 2017
M. Turzi, *The Political Economy of Agricultural Booms*,
DOI 10.1007/978-3-319-45946-2_5

"quarantined" Paraguay, de facto making the country self-sufficient and the most advanced economy in the Southern Cone. This modernizing "for life" dictator tradition was followed by successor Carlos Antonio López (1844–1862) and later by son Francisco Solano López (1862–1869). Francisco parted from the autarkic principle and attempted to expand Paraguay's role in regional affairs, sparking a war with Argentina, Brazil, and Uruguay. After the five-year *Guerra Grande* (great war, 1864–1870), the population was decimated (50 % of the total population and 90 % of the males had perished), the country was defeated, the territory dismembered and occupied by Brazil, and state finances burdened due to the imposition of heavy compensation. In order to meet these obligations, the government began to sell land. With all domestic capital consumed in the war, foreigners bought vast tracts of a country in which, until 1870, close to 90 % of land was publicly owned. Without any consultation to the indigenous residents of those lands, companies like the Anglo-Argentine[2] *Industrial Paraguaya*, bought 2.6 million hectares in the country. Paraguay suffered the most violent Brazilian military incursion ever, but it remained under Brazilian sub-imperialist expansion through soybean production (Oliveira 2016). Brazilian tutelage since 1870 anchored the power of Paraguayan agrarian ruling elites. In the process, it also consolidated the dependent position of the political system. The Colorado party—the main organized political force in the country—ruled under Brazilian guidance between 1887 and 1904. Power sharing agreements were never implemented, and the strongman *caudillo* patriarchy structured Paraguayan political culture.

The latest episode in this tradition was Alfredo Stroessner's forty-four year-long dictatorship (1954–1989). With the administrative support of the Colorados, he consolidated personal authority over the institutions of the state as the *pater familiae* of the Paraguayan political system (Roett and Sacks, 1991). The arrangement was possible due to a functional division of labor: the Colorado party ran the government and bureaucracy; the military was in charge of repressing (the opposition and dissenters, mainly in the first years); cronyism; and rampant corruption (clientelism and bribes) glued allegiances. In the rural areas in particular, the *stronista* regime never perceived peasants or indigenous populations as a political or military threat. With the vast majority of them speaking Guaraní and living from subsistence farming, they were invisible in political terms. Clashes between small holders and big landowners have been recurrent in the Paraguayan rural environment. But when the economy expanded in the 1960s and 1970s, the agricultural frontier opened up, and there was a developmental drive to transform the Chaco region into cattle ranches and the forests into cotton fields. According to Nickson (1981), the government launched a program to increase production, relieve population pressure, and encourage agricultural modernization. To organize this process, the Institute of Rural Welfare (IBR) was created in 1963. Its main task was to remove—the accepted word was "relocate"—squatters and poor farmers in new agricultural colonies in the north and eastern regions, a longstanding demand of the *latifundistas*. The expansion of the agricultural frontier brought about the forced resettling of local farmers and peasants, the majority of which had a precarious legal hold on their lands or were simply squatting. To secure his hold on the country, Stroessner distributed the country's land to co-opt the political elite and keep the military loyal. However, they sold it to Brazilians and American companies instead of becoming producers, which would have consolidated a national landed elite. Conflicts over boundaries and communal grazing rights led to increasingly louder outcry about the injustice of the existing land tenure system. In the early 1970s, Church-sponsored *Ligas Agrarias* (land leagues) started to emerge. Cross-border commercial soybean plantation during the 1970s was facilitated by the close geopolitical ties between the military dictatorships in Paraguay and Brazil. The productive economic situation would be politically institutionalized in the *Friendship and Cooperation Treaty'* of 1975, which protected the presence of *brasiguayos* and Brazilian farmers in eastern Paraguay.

In Paraguay, soybean production created a new class of agricultural barons. It redistributed power within the rural structure; still keeping it concentrated in land, production, and power terms. The Brazilian institutional response was political adaptation coordination. In Paraguay, the economic explosion has reinforced the existing social structure and fortified the position of the previously dominant actors. The main area of conflict in Paraguayan soybean production also revolves around land distribution, aggravated by the fact that the haves are foreign, *brasiguayo*[3] producers and the have-nots are indigenous Paraguayan peasants. FAO's world census of agriculture ranks the country as the most unequal in land distribution in the world—second only to Barbados—with a Gini index of land concentration of 0.93. The latest National Agricultural Census conducted by the Ministry of Agriculture (MAG) in 2008 revealed total agricultural area increased 36 % (23.8 M to 32.5 M ha) from the previous census. Still, 85.5 % of that area is owned by 2.6 % of producers. For the soybean tract, total cultivated area increased 345 %. Confirming the productive concentration inherent to its production model, units of <20 ha decreased 11 % while units in the >1000 ha range soared 1701 %. While only 16 % of producers own >100 ha, they account for 87 % of cultivated area. At the same time, 66 % of soybean producers have <20 ha, only amount to 4 % of cultivated surface. In 1991, there were 26 soybean producers with over 1000 ha. By 2014 there were 482—a 1,753 % increase. Because of the productive concentration inherent to the soybean model of production, the political economy of production has a vested interest in perpetuating Paraguayan land concentration. According to the 2002 census and interview-based evidence, *brasiguayos* control the majority of the extensions between 50 and 200 ha. The >500 ha segment is also dominated by *brasiguayos* and their Brazilian partners. In a country with 92 % of its exports originating in the agricultural sector, control of the fastest growing crop has a critical impact in the governing institutional configuration. The impact is only magnified if this product is the source of wealth from which to generate a redistribution push to lift the 22.6 % of the population that lies below the national poverty line. Paraguayan population on an income below $2.5 a day was 12.67 %, according to the Inter-American Development Bank latest data.

The Paraguayan case can be classified as *colonization*, understood as the establishment of non-native settlers who carry out productive activities in a territory they claim as their own.[4] This definition may seem narrow and deliberately neutral, but it helps to circumvent the politically charged concept of colonialism, which emphasizes intention, and focuses on the ruling of new territories and existing peoples. The main cause of this skewed distribution is chronic state fragility, which has shaped the process of colonization. As understood here, this concept has two dimensions:

- The first one is *domestic* and refers to the colonization of the Paraguayan state by private agrarian interests (state capture), which has created an immense distributional gap in the countryside between landlords and landless. Domestically, the state in rural areas (police, magistrates) is indeed—as Marx would have posited—an instrument of the (landed) capitalist class. The soybean boom has (re)empowered the actors with vested interests in keeping institutions with low levels of *public regardedness.*[5]
- The second one is *international*, associated to Paraguay's position in the Southern Cone. As part of the Brazilian hinterland, soybean production in Paraguay was in its inception—and still remains

today—dominated by the *brasiguayos*. The link connecting the *brasiguayos*—as a social force and economic group—to the Brazilian state is the agribusiness Brazilian sector. In 2008, Lula enacted decree 6.592 to regulate the National Mobilization System, which confronts "foreign aggression."[6] Increasingly, militant agitation by landless peasants and indigenous peoples in Paraguay for redistribution of Brazilian-owned soybean farms produce an alignment between geopolitical imperatives and agribusiness interests. The signal from the Brazilian state was clear: expropriation would produce strong diplomatic, economic, and military reactions. Paraguay's position between more powerful Argentina and Brazil meant a constant vigilance and a heightened sense of resistance to domination by its neighbors. Geopolitics, which for many years was the dominant regional foreign policy framework, would dictate Paraguay to be fated to a submissive role.

THE BRASIGUAYOS: AN INTERMESTIC DRIVING FORCE

Soybeans have constituted the springboard for Brazilian land ownership in Paraguay. Brazilian migration to Paraguay agricultural lands was not a spillover of the 1950s and 1960s process of Brazil's expansion of its own agricultural frontier, or merely the economic response to production cost differentials. The strengthening of bilateral ties also reassured Brazil that its Paraguayan periphery would not become a "turbulent frontier."[7] For Paraguay, it was a foreign policy tool to forge closer ties with Brazil. Both countries found the rapprochement instrumental to a geopolitical design in which Paraguay could reduce its dependence on Argentina and Brazil could weaken it. Stroessner gravitated toward Brazil by establishing transportation links to the west. Brazil granted Paraguay free-port privileges on the Brazilian coast at Paranaguá and built the Amistad bridge over the Paraná river. The Treaty of Itaipú in April 1973 symbolized the highest point in bilateral ties. The *stronista* regime benefited both politically and economically from its association to Brazil, which in turn strengthened Stroessner's domestic position.[8]

Neupert (1991) describes a process of "intense colonization" of the Paraguayan agricultural frontier from the early 1960s to the mid-1980s. The Brazilians were pre-eminent, although other groups such as Japanese, Mennonites, or even the Korean Moon sect[9] were present. Many *brasiguayos* were able to buy extensive tracts of land for export crops on small- and medium-sized holdings.[10] Small farmers were displaced by mechanized agriculture in the Brazilian southern states of Paraná, Santa Catarina, and Rio Grande do Sul. As a result, they were attracted to Paraguay by its proximity, growing economic prospects, fast infrastructure improvements, availability of land at low prices, and favorable credit and tax policies. However, the soybean boom was not happening in a vacuum. Pre-existing land concentration patterns in Paraguay perpetuated the exclusion of around 75 % of rural workers from ownership. This situation was very different from what happened in neighboring Argentina, where conditions and objectives of rural colonization in 1870–1910, coupled with the low population density, resulted in an atomized rural structure very similar to the one experienced by the US farmers. As previously mentioned, conditions in Brazil were more similar to the Paraguayan case than to the Argentine. The Stroessner government wanted to create a wave of settlers to occupy the Chaco for national security purposes. As part of this effort, Brazilian migrants even enjoyed better conditions than Paraguayans on Paraguayan soil. Nickson (1981: 119–120) documented how the *brasiguayos* were able to obtain much softer loan terms with low interest rates[11] for agricultural development from the State Development Bank (BNF), the IDB, and the World Bank. These initial conditions excluded Paraguayan farmers from using land title as collateral, closing any possibility of obtaining credit to fund technological improvements. To make matters worse, this technological component was becoming increasingly important to maintain competitiveness in soybean production. Indeed, technology was becoming a fundamental element of production, without which small farmers were driven out. In practice, the lack of credit created impassable barriers of entry for local peasants and small farmers vis-à-vis Brazilian migrants.

Coupled with total absence of financial aid, technical assistance or training programs,[12] concentration became a self-supporting circle in which the *brasiguayos* have ended up owning more land with the profits they obtained from the enhanced competitiveness derived from previous investments. Although conclusive figures are lacking, thousands[13] of *brasiguayos* have transformed the Paraguayan rural areas into export-oriented sites of soybeans and cotton production. Today, of the 17 departments in the country, all but five report a significant presence of *brasiguayos*.[14] In the major soybean-producing departments of Canindeyú, Alto Paraná, Itapúa, Caagazú, and San Pedro, production is dominated by the brasiguayos. Tranquilo Fávero is considered the biggest soybean producer in Paraguay, with 55,000 ha in 13 different departments and over 30 silos. However, there is no nationalist backlash against what constitutes at best economic colonization. Both *brasiguayo* and Brazilian producers have integrated with the "national agricultural bourgeoisie"[15] and the big Paraguayan landowners. The latter are mostly former generals from the stronista period who benefited from the fraudulent distribution carried out by General Stroessner. The interest cluster is completed by international seed companies,[16] traders/exporters[17] and agribusiness groups[18] who also operate in Paraguay as a source of cheap beans for the Argentine oil industry.[19] The Paraguayan Chamber of Grains and Oilseeds Exporters (CAPECO), the Soybean Producers Association (APS), and the powerful Farmer's Union Syndicate (UGP) and Paraguayan Agricultural Coordinate (CAP) blend legislative lobby with direct action in the streets or *tractorazos*, tractor and agricultural machinery displays and roadblocks. As explained by president of the Commission for the Development of Yguazú Ichiro Fukui: "*We demand to be included in the Agrarian Reform Council. We are not against the government, we just want peace in order to work and bring progress, to bring progress to Paraguay.*"[20]

Then executive director of Oxfam International, Jeremy Hobbs, said in 2012 to the New York Times the *brasiguayo* expansion was responsible for a prolonged conflict over what locals call "earth robbery." For the period 1992–2012, 100,000 small-scale local farmers have either migrated to city slums or abroad or have become landless. Each year in Paraguay 9,000 rural families are evicted by soy production and nearly half a million hectares of land are turned into soybean. Conflict was also ignited with indigenous groups who were expelled from their lands and relegated to the edges of society, attempting to force their assimilation into the mass of the rural peasantry. After a brief initial resistance, by the 1980s and 1990s the situation had changed. The new governmental strategy was differentiation (Horst 2007), which led to an exclusion of the *indigenista* movement from national political representation.[21] Paraguayan indigenous peoples have not unified their claims with the peasant movement, although their grievances stem from the same status quo of land distribution and tenure. Peasant organizations are the ones at the forefront of the struggle against the soybean model of production. They operate at the district, department, and national level. At this last level, the National Coordinator of Peasant Organizations (MCNOC) assembles the Paraguayan Peasant Movement (MCP), the Fight for Land Organization (OLT), the National Peasant Union (UCN), the National Independent Aboriginal Organization

(ONAI), the Worker Peasant Front (FOC), and the National Peasant Federation (FNC). However, the organizational strength is in the districts, where local peasant leaders—who know the players, their interests, and power resources—are decisive for action. The peasant movements have heterogeneous ideologies and party affiliations, but they coincide in taking their claims outside the institutional arenas and into direct action: roadblocks, demonstrations in front of the estates to prevent sowing or avoid agrochemical use, and invasion are the most common means of protest. The aims are to "hold back the Brazilian dominated soybean model of production," which, in their view, destroys their livelihoods by displacing populations, taking away their land, closing work opportunities, and contaminating their crops, livestock, and families. In comparative perspective, Brazil has been much more successful at organizing resistance from below, with peasant and indigenous movements creating "political scales," unifying their platforms and thus overcoming organization costs. Qualitative studies (Fogel 1989; Fogel et. al. 2005) point to internal divisions among ethnic groups as the main impediment for this coalition building to happen in Paraguay. It is also indicative of a much more powerful position of the dominant actors in the chain and of much more rigid control over the means of production, which creates a social structure in which relations of production are more hierarchically organized.

Although there are no official figures, estimates of brasiguayo control of the soybean chain range between 60 % and 80 %, depending on whether Brazilian partners are included or not. This figure includes not only the land with the grain, but also the related linkages such as storage and transport. Taking into account that a substantial part of the soybean chain—namely, input provision and trading—is also in the hands of foreign companies, there is a substantial gain from soybean production that is not accruing to Paraguayans.[22] Conflicts of interest are framed in a logic that opposes the Brazilian modern agribusiness model of strong work ethics and high productivity to a backward subsistence model of family agriculture carried out by lazy Paraguayan peasants. The prevailing discourse is that the foreign minority represents progress and the landless peasants are slothful and backward.[23] Understanding discourse as a contentious social narrative that is imposed by the powerful (Foucault 1977), this kind of evidence confirms Paraguay as an instance of colonization. Brazilian producers have colonized soybean production in Paraguay. This has happened with explicit Paraguayan support, like when President Horacio Cartes encouraged Brazilian investors in 2014 to "use and abuse Paraguay."[24] The conditions are polarized economic differences and inequalities, escalating levels of conflict and—since the downfall of President Lugo—institutional weakness and political regression. The concentration of land and productive activities indicates a model that privileges the interests of the landed elite, agribusiness, and—more broadly—international capital applied to agricultural production (Fogel 2008, Galeano 2012).

The Far West

Land distribution was based on a network of cronies and associates under the pretense of the 1963 agrarian reform program. Using data from the Institute of Rural Welfare (IBR) and of the Land and Rural Development Institute (INDERT), Alegre Sasiain et al. (2008) has proven that many of the subjects of the agrarian reform—who in such capacity were granted land—were in fact the most powerful people in the regime, today among the richest people in the country.[25] Even after the fall of Stroessner, the balance of power was safeguarded by the survival of the governing structure (Roett 1989). Lack of institutional density and of public-regardedness of the Paraguayan state agencies have contributed to a "lawlessness" environment. The public administration in Paraguay is highly politicized and inefficient. Clientelistic networks trump formal institutions, especially at regional and the local levels. Finally, the judiciary is highly politicized and corrupt. It functions as part of a balance of power system with the executive and legislative.

Until the Fernando Lugo election in 2008—which ended **sixty-one** years of dominance—the Colorado party dominated the middle and lower echelons of civil service. Landless peasants armed with machetes and clubs regularly attempt to impede sowing, while armed landlord groups and police guard their machinery and inputs. There is a permanent tension between landowners and groups that often camp outside their estates waiting to enter and occupy land, a tension that is more often than not resolved at gunpoint. In September 2009, nineteen-year-old activist Abraham Sánchez Gayoso was shot dead by sentries of the 5000 ha *Iriarte Cué* estate (Urundey colony, Unión district, San Pedro department) after a group of 90 OLT members invaded the premises.[26] Owner Elpidio Rojas had been invaded twice before by OLT and had filed over 17 claims for attacks on workers, usurpation attempts and other desecrations of property. He was frustrated time and again asking police for protection in the wake of imminent[27] invasions. The absence of institutional mediation forces conflicts to be resolved directly by the parties involved, exacerbating violence. Rojas had prefigured the tragedy when he said: "*These so-called landless do not let us work. I bought this land two years ago and we have endured permanent harassment. This makes me very bitter, but next time these andaí[28]* (lazy) *come here, we will have to defend ourselves*". Such incidents are not isolated. San Pedro has more than 50 landless commissions with active land claims before the INDERT.

To complicate matters, since 2008 the leftist Paraguayan People's Army (Ejército Popular dle Paraguay, EPP) has been active in the northern departments of Paraguay bordering the Brazilian border. The EPP proclaims itself as dedicated to a socialist revolution in Paraguay. Membership is believed to range from 20 to 100 members and structure to be small and decentralized. Location has been pinned to the Concepción Department. Activities include kidnappings, placement of explosive devices, shootouts with the police and military, attacks on ranchers and on peasants accused of collaborating with security services. The frontier between land claims and terrorism quickly blurred. On August 31, 2009, MCNOC and OLT agreed to launch a massive occupation plan for ten properties in the Capi'ibary, Choré and Unión districts in San Pedro. OLT Adolfo Villagra had informed the targets were to be *Iriarte Cué*, 2,300 ha; *Lucero SA* in the Santa Catalina colony, 12,000 ha; *Don Pedro* in Capi'ibary, 15,000 ha; *Carla María* (property of CAPECO Director José Bogarín Acosta), 5,300 ha; and *Mbery* in Yryvucuá.[29] By January 2010, after INDERT promised to speed up the process of land redistribution, MCNOC and OLT agreed to put all invasions on hold. Former catholic bishop Fernando Lugo had campaigned as a "champion of the poor," on the promise to redistribute land and carry out agrarian reform. Lugo was elected as the head of a coalition. Vice-president Federico Franco belonged to the conservative Liberal Party. Like the Colorado Party, Liberal Party had close ties to agribusiness. With no political party of his own, and the Liberals and Colorados in firm control of Congress, Lugo lacked political capital to push through agrarian reform. Instead, he responded to campesino complaints by tightening regulation of soybean production. Pushback by agribusiness began almost immediately. The Union of Production Trades (UGP)—an umbrella group that represents Paraguayan companies as well as multinational corporations—and the media[30] attacked Lugo for the continued occupation of land by *campesinos* (peasants). A group of landless peasants occupied in June a 2,000-ha property in the border region of

Curuguaty. Campo Morumbí—a company owned by wealthy Colorado Party senator Blas Riquelme—controlled the land. In June 2012, 324 police officers, a helicopter, and Paraguay's elite anti-terrorism Special Forces units were deployed. After a firefight erupted, 11 peasants and 6 security officers were killed. It took 10 hours for the UGP to demand Lugo's impeachment, a call echoed by media such as ABC Color. Using this as a pretense, Paraguay's Senate impeached Lugo a week later, 39 votes to 4.

Lack of governmental enforcement of the law can be attributed not only to incapacity, but also to complicity. In rural Paraguay, the state turns a blind eye to landowners' criminal shows of force. The subordination of the state is evident in the soybean sowing period, when state institutions such as police and the army go heavily armed into the fields with the brasiguayos in order to secure production against peasant crowds who stand with sticks and machetes ready to occupy a tract of land that they contend belongs to them. In this "state of nature," the outcome is dictated by the more powerful landowners. Not only do captured institutions fail to mitigate these asymmetric realities, they perpetuate them. This section has not delved into the legal dispositions or framework carried out in Asunción for precisely this reason. Enforcement of the capital's norms is trumped by local—real—power. In general, enforcement capacity is inversely proportional to the distance from the capital. As an example, payment of property taxes is spontaneous and voluntary—the producer goes to the municipality and declares how many hectares he possesses and then he is told how much he has to pay.

Agrochemical use is out of control, even in the presence of legal provisions like mandatory barriers for spraying chemicals near trails or villages[31] or a prohibition on cultivation within a 100 m radius from schools, centers of collective attendance (plazas, sport fields, and churches). The most renowned case of this negligence was the death of an eleven year-old Silvino Ramón Talavera, who died after being sprayed with agrochemicals. The groceries he was carrying were also contaminated, making 22 members of his family sick as well. To date, it remains the one case in which producers—Alfredo Lauro Laustenlager and Hernán Schelender Thiebeaud—were sentenced for "manslaughter and production of risk." In February 2009, the Environment Secretary confirmed that the school in Kuñataî, Aba'i district, Caazapá department was blocked by soybean plants. Brasiguayo producer Mauri Karn admitted responsibility for having sown the beans, but attempted a defense arguing *"profits would go to help the school."*[32] A mirror situation exists for production and export. Much of the Paraguayan soybean is part of the contraband trade with Brazil or Argentina. Customs and border patrol officers are easily corruptible. But corruption in Paraguay has become institutionalized and systemic in nature. Transparency International's 2015 Corruption Perceptions Index ranks Paraguay 130 out of 167 countries, the second most corrupt country in South America (Venezuela is in the 158th position).[33]

With regards to land tenure, no mechanism exists to adjudicate the legitimacy of claims either. Landless peasants are sometimes not even invading; for they claim those lands belong to them by law in the first place. Other organized groups occupy land to later sell their *derecheras* (concession certificate[34]). Unchallenged local power means there are no safeguards against abuses stemming from rural social stratification, like the organization of private militias or the respect of labor contracts or human rights. The case of Paraguay exhibits as its dominant feature a subordination of the political system to a specific socioeconomic foreign group. The roots of this colonization are historical (destroyed by a crippling war against its neighbors, Paraguay was occupied and run by Brazil thereafter), political (Paraguay's land concentration pattern emerged during Stroessner's neopatrimonial regime and predatory state (Robinson, 1999), its geopolitical position wedged between Argentina and Brazil), and also economic (the successful capture of state institutions by the dominant player riding the soybean boom). The expansion of the soybean chain has meant growing territorial control by sector interests and the deepening of institutional weaknesses. Economically, it has resulted in a model of low-added-value production, transnationalized and dependent, efficient but exclusionary and with growing enclave features.[35] Politically, it showcases incapacity at best, though evidence from the patrimonial *stronista* regime onwards points to a collusion that has blocked more inclusive growth and hampered state capacity. After being captured by private landed interests, neither central nor local-level institutions have been able or willing to exert control over the process. In April 2013, Paraguay elected rightwing tobacco tycoon Horacio Cartes as president. Despite having been investigated for fraud and drug smuggling, Cartes recorded a clear-cut victory that marks the resurgence of the Colorado party. It also signals the return of a political economy coalition that represents landowners and agribusiness in a country in which agriculture represents over 20 % of GDP.

NOTES

1. *Correo de la UNESCO*, August 1977, 56–59.
2. The company was owned by Argentine Carlos Casado and funded by English capital. There were others, like Brazilian Mate Larangeira-Mendes and Domingo Barthe. Interestingly, not even the 1963 decree by President Stroessner could expropriate a single hectare from Casado's property. To this day, these companies have a pre-eminent position in the Paraguayan agricultural sector.
3. *Brasiguayo* (Spanish) or *brasiguaio* (Portuguese) refers to Brazilian settlers in Paraguay and their descendants. It can also be a person who is of both Paraguayan and Brazilian parents and a Brazilian who spent many years or has many interests in Paraguay. The fluidity of the concept is reflective of the fluidity of the borders and interests between both countries.
4. "Colonization and Colonialism, History of", *International Encyclopedia of the Social & Behavioral Sciences* (2004), 2240–2245.
5. "Public-regardedness refers to the extent to which policies produced by a given system promote the general welfare and resemble public goods (that is, are public-regarding) or tend to funnel private benefits to certain individuals, factions, or regions in the form of projects with concentrated benefits, subsidies, or tax loopholes (that is, are private-regarding). This dimension is closely tied to inequality, particularly since those favored by private-regarding policies tend to be the members of the elite, who have the economic and political clout to skew policy decisions in their favor". Scartascini et al. (2008a: 10)
6. The decree defines (Art. 2, §1°) foreign aggression as "*threats or injurious acts that harm national sovereignty, territorial integrity, the Brazilian people, or national institutions, even when they do not constitute an invasion of national territory*". http://www.planalto.gov.br/ccivil_03/_ato2007-2010/2008/Decreto/D6592.htm
7. John. S. Galbraith, "The turbulent Frontier" as a Factor in British Expansion", *Comparative Studies in Society and History*, Vol. 2, Jan 1960: 150–168.
8. Even in the presence of starkly non-reciprocal conditions. For example, in 1979, Brazil passed law 6.634/79, which established a 150 km buffer zone from its borders, within which no foreigner is allowed to buy rural properties.
9. In the year 2000, the Unification Church headed by Korean Reverend Sung Myung Moon paid US$24 million for 6,000,000 hectares of land from the heirs of Carlos Casado (Alto Paraguay region) for agricultural activities of "La Victoria" company.
10. Producer Virgilio Moreira, who arrived in Paraguay in the 1971, remembered that in those times, selling a soybean hectare in native Paraná state

yielded the price of 3–4 ha of Paraguayan land. Fogel and Riquelme (2005) reported that in Río Grande do Sul, Brazil, a soybean producing hectare costs US$2,500, while in Paraguay an equivalent tract is US$1,000.

11. For farmers in Brazil, the prevailing rate was 22–24 % annual for five year period with a one-year grace period, while in Paraguay the BNF agricultural loan rate was 13 % over eight years and a three-year grace period.

12. Neupert (1991), (Laino 1977), and Fogel and Riquelme (2005) agree that the lack of structural reforms evinces that it was never the Paraguayan state's intention to level the playing field for small farmers or go beyond simple spatial redistribution of the population.

13. The 2008 Agricultural Census of the Ministry of Agriculture finds 8,954 rural producers of Brazilian nationality and 267,180 Paraguayans. Rural Welfare Institute (Instituto de Bienestar Rural—IBR) sources estimate 350,000–400,000. Migrations officers say there are 118,000 legally inscribed immigrants, while Brazilian consular figures are between 400,000 and 500,000 out of a total Paraguayan population of 6.3 million people. Brazilian historian Marta Izabel Schneider Fiorentin in her 2012 "Imigracao Brasil-Paraguai" book puts the figure at more than 500,000, also stressing that the exact figure cannot be determined.

14. Presidente Hayes, Central, Cordillera, Paraguarí and Guairá.

15. *Campesinos* are empowered—symbolically—when the antagonism is framed in terms of nationality. Then, national kinship takes over and reframes the struggle as foreign encroaching landowners vs. citizen-peasants.

16. Monsanto and Syngenta, operating directly or through subsidiaries. Also, Brazilian Embrapa and argentines Don Mario, Relmó and Nidera.

17. Cargill, ADM, Dreyfus, Noble, and Bunge

18. The biggest Argentine players in the Paraguayan soybean market are Carlos Casado (who already owned large soybean extensions in Paraguay) and Cresud. They also allied and created joint venture *Cresca*. Other players include Grupo Los Grobo (through Tierra Roja), El Tejar, and Pérez Companc. The largest Brazilians player is the Grupo Espiritu Santo. An interesting example of local agribusiness group is Kimex SRL, owned by the Kress group.

19. Argentine oil company Vicentín buys most of Grupo Espiritu Santo's Paraguayan production. However, imports were damaged on April 6, 2009, when the Argentine government eliminated the import benefit on Paraguayan soybeans which excluded importers from paying value-added and earnings taxes.

20. On occasion of the December 16, 2008 *tractorazo*.

21. Since Rodríguez de Francia forced the colonial elite to intermarry, the very concept of indigenous identity is fuzzy in Paraguay. Differences between indigenous groups are often greater than the ones between natives—if that category can be properly applied—and the rest of society. To make matters

worse, there is no indigenous collective consciousness component in the national creed, as is the case in Bolivia.

22. As an anecdotal example, machinery used in brasiguayo or Brazilian controlled fields is made entirely in Brazil. It is brought—many have argued illegally—into Paraguay and returned afterwards.

23. Albuquerque, José; *Campesinos paraguayos y "brasiguayos" en la frontera este del Paraguay*, in Fogel and Riquelme (2005).

24. www.ultimahora.com/cartes-empresarios-brasilenos-usen-y-abusen-paraguay-n767800.html.

25. Among the "VIP" list are: Stroessner's private secretary Mario Abdo Benítez, Chief of intelligence Pastor Coronel, Defense Minister Marcial Samaniego, Itaipú director Enzo Cesare Debernardi, Vice President Luis María Argaña, Public Works and Communications Minister José Alberto Planás, current Senator Bader Rachid Lichi, current Judge Wildo Rienzi Galeano, Stroessner's son's lawyer Hirán Delgado Von Lepel, Paraguayan Workers Confederation leader Sotero Ledesma, Finance Minister César Barrientos, former Colorado Party president Eugenio Sanabria Cantero and Generals Alejandro Fretes Dávalos, Otello Carpinelli, Enrique Duarte Alder, Humberto Garcete, Orlando Machuca Vargas, Gerardo Alberto Johansen and Roberto Knopfelmacher.

26. District Attorney Rosa Talavera is yet to determine if it was a skirmish gone awry or an ambush, but she ordered the detention of *capataz* Leoncio Esquivel Domingo Fernández, José de Jesús Vallejos and Venancio Fernández. After the occurrence, Rojas told radio 780 AM he was sorry his employees were jailed and he appealed for their release on the grounds they were workers and not hired guards. He accused Talavera of being partial and committing the "huge injustice" of detaining his employees and not the peasants under usurpation charges.

27. Landless movements announcewhich of the estates they plan to occupy. Targets are chosen on the basis of the claim that landlords have idle land or that they have not offered satisfactory proof of rightful ownership.

28. This "pioneer" discourse repeats the colonialist civilizing mission, this time around emphasizing work ethics. What in their own country plays out as a frontier mentality, in Paraguay is recreated as a "Brazilian man's burden".

29. "Campesinos iniciarán hoy las ocupaciones masivass", *ABC Digital*, August 31, 2009. MCNOC had targeted the Choré district, where it laid claims to 1.510 ha. *La Solución SA*, and to 1.000 ha. *La Fortuna, La Palomita* and *Agroganadera Jejuí*.

30. "It's the left of Hugo Chavez, the left of the Castros, a left that fanaticised Paraguayan campesinos," said Aldo Zuccolillo; owner of ABC Color, Paraguay's largest and most influential newspaper.

31. In May 2009, Congress passed Law N° 3742/2009, easing regulation (which had been upgraded in April in Decree 1937/2009) on agrochemical use. The Law was vetoed partially in July by Decree N° 2361/2009.
32. "Cultivan soja hasta en el predio de una escuela en el distrito de Aba'I", *ABC Digital*, February 05, 2009.
33. Even with the $5.5 billion *Petrolão* scandal, Brazil appears ranked 76.
34. Granted by the IBR, each *derechera* corresponds to 7–10 ha and can be paid in five years, after which property title is granted. But this last process can take up to ten years.
35. In such economic organizations of production, the benefits remain confined to an international sector not connected to the wider economy. Linkages are few and weak, as are the distributional effects.

CHAPTER 6

Confrontation (. . . and Beyond): Argentina

Abstract Argentina is a case of centralized institutions exhibiting a conflictive pattern of relations with the economic sector/resource (confrontation). The soybean agricultural model of production is inherently opposed
to the governing Argentine political economy configuration during that
period: the labor/industrial/urban coalition on which the Peronist party
built its ascendancy in Argentine politics. The three Kirchner governments
(2003–2015) display a domineering and confrontational political style.
Despite the recent governmental change—the election of Mauricio Macri
in 2015—there is a built-in institutional imperative to exact rents still
remains, and the agricultural sector remains a prime target.

Keywords Agriculture · Agribusiness · International political economy ·
Latin America · Soybeans · Argentina · Kirchner · Macri

The Argentine soybean chain is not besieged by the conflicts of its neighbors: access to land and property rights are not contested, and it does not
have an Amazonian frontier to deal with as a major ecological component
of agrarian policy. As no major groups (landless peasants, indigenous
communities) dispute the mode of production, the rural social structure
is much less polarized, with no dramatic escalations or the recurring armed
standoffs observable in Brazil and Paraguay. However, the relationship
between the agricultural sector and the government has been anything but

M. Turzi, *The Political Economy of Agricultural Booms*,
DOI 10.1007/978-3-319-45946-2_6

117

harmonious. The previous sections established how the Brazilian soybean chain integrated with a political system *responsive* to its interests and Paraguayan sectoral interests overwhelmed an incompletely democratized *colonized* state. In the Argentine case, a *confrontational* relationship existed for the period 2000–2015 between the state and the productive agricultural sector. This situation stems from the fact that the soybean agricultural model of production is inherently opposed to the governing Argentine political economy configuration during that period: the labor/industrial/urban coalition on which the *Peronist* party built its ascendancy in Argentine politics. The main line of conflict in Argentina is thus not endogenous to the sector, but between the sector and the governing structure.

A State Against the Campo?

As a political economy paradigm, the agro-export model behind the idea of Argentina as the world's breadbasket was challenged after the World War II. The international context triggered by the Great Depression called for autarkic responses. Declining terms of trade for agricultural products and—later on—industrialized countries' subsidization of primary production acted as further disincentives for this model. As a result, countries in Latin America sought to industrialize via import substitution (ISI). Since economic processes cannot be separated from political ones (Ordeshook 1990), this developmental model had its political coalition correlate. A "labor mobilizing" (Murillo 2008, Murillo and Schrank, 2009) strategy consolidated a political economy structure that has been alive since the mid-1940s: an urban/industrial coalition of workers and preferred business arbitrated by Perón himself at first and the Peronist party later. Labor-backed parties in government would enact barriers, tariffs, quotas, and other regulations to protect state-owned or publicly subsidized companies. These policies benefited the national corporate sector and the unions, who with job stability gained members and influence. In return, labor translated its affiliates into electoral majorities. For the past sixty years, the pillars of this corporatist model have remained, though they have morphed into a "segmented neocorporatism" (Etchemnedy and Collier 2007; Mc Guire 1997). In practice, workers' welfare has been engulfed by the unions' bureaucracy.[1] The national bourgeoisie has survived by state favors rather than by gains in efficiency, and the line between the Peronist party and the state apparatus are seldom distinguishable. This dominant political economy configuration, especially strong after the 2001–2002 crisis, has remained paramount in the struggle between multiple distributional demands. Capitalizing on the collapse, industrialists pushed for the devaluation of the peso—pegged to the dollar under the convertibility scheme—to swing the pendulum back toward national industry. Politically, Argentina had a longstanding tradition of fiscal centralization and concentration of political power in the Executive branch, an inheritance of the historic battle of Buenos Aires against the *interior*. The richest province, Buenos Aires controlled the ports and customs office. Even though the provinces declared their independence from Spain in 1810, Buenos Aires proclaimed itself an autonomous political entity in 1820. Between 1830 and 1853, notably under the governorships of Juan Manuel de Rosas (1829–1832 and 1835–1852), Buenos Aires was pre-eminent to the rest of the country, which was not yet unified. Indeed, it was not until Rosas' fall that the nation came together as one under the 1853 Constitution. Yet, Buenos Aires resisted and seceded. After the decisive victory at the battle of Pavón in 1861 the civil war ended and Buenos Aires was incorporated into the Republic. Institutional "path dependence" (North 1990) established that control over resources remain crucial to sustain governors' allegiances; they are the territorial bosses that provide the "basic units" of power. Both at the state and the Peronist PJ (*Partido Justicialista*) level, governing capability depends on effective clientelist exchanges. These exchanges are the beating heart of Argentine federalism. Modern patronage networks resemble the Avellaneda/Roca centralist system of control known as the "league of governors" (1870s), which tied the president to the leaders in the provinces, creating a de facto consortium of provincial political leaders (Rock 2002: 56). Governors, congressmen, or even officials do not effectively represent the interests of their constituency because they are politically obliged to the PJ party structure and depend financially on the Executive.

The underpinnings of the hyper-presidential corporatist state created institutional biases against the agricultural sector. The urban/industrial coalition is based on supporting a sector with low levels of productivity (Krugman and Obstfeld 2003) but highly organized interest representation and high capability of exerting political and electoral pressure (Isern Munne 2007). The reverse situation happens in the agricultural sector. Export products are internationally competitive, accounting for more than 55 % of total Argentine exports, with soybean complex exports alone representing 28 % of total exports and 34 % of all complex exports, according to governmental 2015 data.[2] The sector is able to produce a sizeable amount of revenues. Besides this capacity to provide funds, the agricultural sector is characterized by historical political fragmentation and low political pressure capacity. Organizational costs are much higher in the agricultural sector, and coordination much more difficult. Mobilization also plays an important role, since it has more marginal benefit in urban than in rural settings. Krueger (1993) introduces a dynamic of "vicious cycles": economic policies such as import protection lead to political economy outcomes. For example, local industrialists make profits not consistent with their competitiveness but with their political influence. The import-competing sector gains influence to the detriment of agricultural exporters, further reinforcing the support for initial policies. The Argentine political economy dynamic is structured around this anti-agriculture bias: a hyper-presidency (Corrales 2002) is supported by an urban/industrial/unionized coalition. The political imperative for the executive's survival is to deliver goods (economic transferences, infrastructure projects, or social plans) to this coalition while keeping the federal political interests at bay. Economically, this means obtaining rents from a sector—like the agricultural—that can provide them steadily, and at the same time using tributes—like ETs—that do not weaken the central government vis-à-vis the provinces. As the result of repeated human interactions (North 1990: 40), institutions have reflected this bias. Institutions processed the conflict displaying all the traits mentioned before. Thus, the economic role of the Argentine agricultural sector has been structurally at odds with the political governing coalition (O'Donnell 1978). The political economy structure of the Argentine case thus has three main characteristics: (a) the urban/industrial coalition is politically effective, despite its economic inefficiency (low incentive from state institutions to extract); (b) the agricultural sector has the capacity to generate resources *and* lacks the political organization to guard them from politically motivated seizure (high incentive from state institutions to extract and ease of implementation); and (c) a centralist power structure demands resource capture for patronage distribution (systemic/structural conditions to extract resources to maintain control).

In this political economy structure, the context of the post-2001 Argentine economic collapse and devaluation led president Duhalde to reintroduce an ET policy. He initially imposed taxes of 10 % on the sales value of major agro-export crops. The taxes were collected from export companies, which passed the burden to producers through depressed purchasing prices. Rates were soon increased to 20 % in search for much needed foreign exchange. During the Néstor Kirchner government (2003–2007), the undervalued exchange rate was intentionally maintained to promote exports, while ETs were a means to subsidize domestic consumption. President Kirchner maintained the ETs throughout his term and increased rates several times. The tax on soybean reached 27.5 % in January 2007 and 35 % in November 2007. Under President Cristina Fernández de Kirchner (2007–), the tax rate became flexible (see the discussion on taxes in Chapter 3 for details). She increased the tax on soy to 44 % shortly after taking office in 2008, triggering the producer protests. ETs two main purposes: extract revenue from the highly profitable agricultural sector and executive control. They could be legislated by decree[3] and were not subject to revenue sharing with provincial governments. Resource centralization was a centerpiece of the couple's strategy to concentrate power by selectively endowing or withholding funds to the governors. ETs became the selected instrument to see this strategy through. Why? Because in the system of revenue sharing between national and provincial governments, ETs belong exclusively to the national government, unlike other fiscal instruments that must be shared with governors. Increasing the resource base through higher ETs strengthened the Executive vis-à-vis provincial interests. Fiscal revenue—by 2013 soybean taxes represented 53 % of total ETs and by 2014 they were 10 % of total fiscal revenue—helped the central government re-establish fiscal solvency after the 2001 crisis and achieve fiscal surplus in the years following, critical for keeping macroeconomic stability and political fidelity. Political loyalty in the Kirchner coalition demanded encouraging import substitution and accessible credit for industry, to improve tax collection via stimulation of consumption. Soybean ETs were a key component of industrial policy. These E's equalized the profitability of primary and processed products, thereby stimulating agroindustry, which was supposed to generate employment.[4] Agricultural products experienced a boom period due to devaluation and the end of the dual exchange rate regime that had in the past crippled agricultural exporters. The devalued peso made Argentine exports cheap and competitive abroad. International agricultural commodity prices were high due to Chinese demand, which injected massive amounts of foreign currency and China becoming a major buyer of Argentina's soybean. Soybeans and derived products generated more export and fiscal revenue than the rest of the whole agricultural chain put together. An incentive to export exists when favorable prices can be obtained in the international commodity markets. However, export booms cause domestic prices to rise, undermining effective purchasing power of urban workers. Export restrictions were intended to mitigate these effects, but instead led to balance-of-payments crises because of resulting trade deficits. Paradoxically, the only way out of these deficits was through export promotion. The end result was a "stop-go" pattern of economic growth: periods of sharp expansion ended abruptly by foreign exchange crises and recessions. Soybeans, not consumed domestically, seemed to have solved this conundrum, for increasing exports did not come at the expense of domestic supply reduction and its consequent increase of domestic prices and real wage reduction. Nevertheless, it produced the same cyclical result of political instability and conflict.

Despite the favorable context, this agricultural export boom sparked the biggest political conflict of the Kirchner era. Why? Just as the political system demands a constant search for revenue sources that can sustain the patronage cycle, political economy constraints prescribe the agricultural sector as the source of those funds. Robinson's (2006) insight is relevant: politicians tend to over-extract (in this case, over-tax) natural resources relative to the efficient extraction rate because they discount the future too much and because resource booms improve the efficiency of the extraction path. The possibility of taxing the agricultural sector (land-based rents) without facing strong opposition, induces politicians to short-term interests. But as the author also cautions, the overall impact of resource booms on the economy depends critically on institutions, since they determine the extent to which political incentives map into institutional constraints. The lack of institutional channels to process conflicts in the sector is not accidental, but rather a reflection of a broader political economy structure that extracts from the agricultural sector to transfer to the industrial. The Inter-American's Bank *producer support estimate* indicator—which reveals what percentage of producer's revenue is due to the support provided by agricultural policies—was in 2012 at – negative 18 %. In comparison, Paraguay in 2013 was at 1.8 % and Brazil in 2014 at 4.4 %. Conversely, the *consumer support estimate*—which indicates how agricultural policies affect the cost of the basket of agricultural products—was for the same countries and years at 13.3 %, 0 %, and –0.3 %. When conflict arose, what happened was what the Brazilian legislators from the *bancada ruralista* fear most[5]: the disagreement had no political mediation and hence the factions grew further apart, transforming the disagreement into an escalating dispute, evidencing the rupture between the political class and the agricultural sector. The state structure only seems to acknowledge the agricultural sector as a source of contributions to the Treasury, while the sector attempted "defensive avoidance" (Lebow 2007) from political intrusion due to its lack of organized political representation. For the agricultural sector, the conflict was a political "Big Bang": it had not been a unified—much less mobilized—political actor prior to 2008. The rural sector's contact with government officials was limited and interaction with institutions uncoordinated, their demands ignored or rebuked. The conflict spawned a new actor in the political arena: the *campo*. Small and medium-sized *mate*-drinking farmers and their families were in the road-blocks (*piquetes*), camping in or along highways and protesting on television, foiling the government's attempt to frame them as "*abundance pickets, the pickets of the wealthiest sectors.*"[6] Representatives closed ranks and formed a Liaison Table (*Mesa de Enlace*), which managed to act as a unified front despite the divergent interests within the sector.[7] Its four main members are: the Rural Society (SRA), historically a cattle ranchers' group, members were also big landowners because of the country's extensive cattle farming mode of production; Argentine Rural Confederation (CRA) represents regional organizations of medium-sized ranchers of the interior, like the powerful Buenos Aires-based CARBAP; the Agrarian Federation (FAA)—a union of small and medium producers—and the Agricultural Cooperative Confederation (CONINAGRO) is an association of rural cooperatives, having a broad network of grassroots organizations. Complementing direct political participation, a less conspicuous lobby structure centered on the technological component of the soybean chain[8] gave logistical supply and reportedly channeled funds[9] to the sector: the Seed Association (ASA), the Agribusiness Chambers Association (ACTA), the Argentine Association of Regional Consortiums for Agricultural Experimentation (AACREA) and the Argentine No-till Farmers Association (AAPRESID), funded by Monsanto to disseminate direct sowing. These efforts were also supported by the Darsecuenta Foundation, an intellectual outlet funded by Bioceres S.A., a consortium of over 190 companies investing in agricultural R&D. The *Mesa* created an institutional mechanism in which the competing agricultural interests of this fractious sector were prioritized and agreed upon. This consensus countered the divide and rule strategy the government was employing. Even against its own particular interests, the *campo* remained cohesive as an actor and established alliances with—in a counterintuitive fashion—the urban upper-middle classes. This alignment cannot be attributed solely to the parties directly involved or gaining from agricultural production. In fact, a governmental defeat was not in the urban upper-middle classes' best interest. Urban dwellers were benefitting from the artificially low prices of transport and major utilities, made possible with funds squeezed out of the campo. Maintaining cohesion allowed for more effective coordinated action, and the sector was strengthened. Fortified and more organized, it has been advancing its interests through the press and through political parties and by lobbying lawmakers. The agricultural sector realized that their lack of participation in the political process had created a vicious circle detrimental to its own interests. Opposition parties incorporated *ruralista* figures into their tickets, capitalizing on the popularity of the *campo* cause. *Agro-deputies* elected in 2009 like Alfredo de Angeli, Pablo Orsolini, Ulises Forte, Estela Garnero, Juan Casañas, Ricardo Buryaile, Hilma Ré, Jorge Chemes, Lucio Aspiazu, Gumersindo Alonso, and Atilio Benedetti had all but one lost their seats.

The new Macri administration was in its first five months in the direction of radically altering the political economy. It named as its first Agroindustry Minister former vice-president of CRA, Ricardo Buryaile and removed or decreased taxes, as shown in Chapter 3. After ET removal, figures from the first trimester of 2016 show agricultural products gained share in exports, while manufacturing lost it. Primary products exports amounted to $1.305bn and shipments of industrial goods $1.25bn. In the new argentine export basket, commodities account for 29.3 % of total exports, against 28.1 % of industrial products. To understand the magnitude of power and wealth redistribution between the different coalitions (rural/agricultural vs. urban/industrial) during the last year of the CFK administration primary goods totaled 20.3 % of total exports and manufacturing accounted for almost double, 39.8 %. The last time exports of primary industry outpaced exports of industrial products—reflective of a new political economy in the country—was in March 1999, when sales of commodities stood at $589 million and shipments of manufactured products at $580 million. Macri assures his "dream is to stop exporting these grains to export branded food, which means workers who will flock to plants and factories to transform wheat into noodles and cookies, corn into breakfast cereal, milk into cheese and so on."[10] The assessment of the incoming administration is that agricultural productivity has been affected by "myopic" and "opportunistic" policies aiming to capture rents from the land dominated by short-term political needs. However, the data indicates otherwise. While soybean exports alone rose 37 %, processed foods and products of regional economies marked a sharp drop in the first three months of 2016. Dairy products exports decreased 37 %, grain mill products (flours and preparations) were down 19 %, fresh fruits dropped 12 % and commercial hides/skins 23 %.

Macri seems to be moving in the 1990s direction in its macroeconomic policies. These in turn define much of the conditions on which the agricultural sector operates, in terms of market signals (exchange rates) and state constraints (taxes). There is also an impact from policies for competing sectors: there is a close relation between structural changes in industry and its effects on agriculture. Finally, the global context—international prices and markets, trading norms and technological availability—sets differential opportunities and restrictions. So far, the government has apparently been seeking to take full advantage of natural resource endowments, exploiting comparative advantages in agribusiness via private investment and trade liberalization. Deregulation and further internationalization of the sector seem to be the guiding principles. In the mentioned decade, this had the effect of raising productivity and increasing production. The overall efficiency gains came in that decade together with crop and land concentration. Argentine agriculture underwent higher degrees of foreign participation and control of the soybean chain. Should these macroeconomic, international, and sectorial realignments happen, there will be political effects that could potentially transform argentine political economy structure and dynamics. The direction can be hypothesized, albeit the success or extent cannot be foreseen in advance.

NOTES

1. The defense of workers' interests is still symbolically important, but residual. While unions do obtain real wage gains and changes in labor laws, more important are organizational and particularistic gains, such as appointments in the state offices (like the ones overseeing the union-controlled health system), and benefits targeted toward more allied unions, like transport subsidies (truck drivers) or appointments in the board of renationalized enterprises (water, mail).

2. National Statistics and Censuses Institute (INDEC), *Export Complexes* indicator classifies argentine exports from a production chains point of view: exports and percentage share of complex exporters (soybean, automotive, oil and petrochemical, gold, steel, copper, fruit and fishing, cotton, ovine, bovine). Available at: http://www.indec.gov.ar/ftp/cuadros/economia/complexp_exportaciones_2010–2014.xls

3. Export taxes allow the president to circumvent a politically costly debate in Congress due to a clause in Law N° 22.415 (Customs Code), which empowers the Executive branch to rule on these tributes.

4. World Bank Data from 2014 estimated agriculture (employment in Argentina to be 1 % of total employment, with 75 % in services and 24 % in industry. This distribution has remained steady on average since 2000. The *agriculture* sector consists of activities in agriculture, hunting, forestry, and fishing, in accordance with division 1 (ISIC 2) or categories A and B (ISIC 3) or category A (ISIC 4). The industry sector consists of mining and quarrying, manufacturing, construction, and public utilities (electricity, gas, and water), in accordance with divisions 2–5 (ISIC 2) or categories C–F (ISIC 3) or categories B–F (ISIC 4). The *services* sector consists of wholesale and retail trade and restaurants and hotels; transport, storage, and communications; financing, insurance, real estate, and business services; and community, social, and personal services, in accordance with divisions 6–9 (ISIC 2) or categories G–Q (ISIC 3) or categories G–U (ISIC 4). The Agrcultural Foundation for Argentine Development (FADA), estimated in 2015 corn generates 1 job every 10 ha, wheat every 20, and soybeans every 50.

5. According to parliamentary Ronaldo Caiado, *op. cit.*

6. Cristina Fernández de Kirchner, *Presidential Address*, March 25, 2008.

7. "*AAPRESID represents only 3000 producers. It is the union of the big sowing pools and mega producers sponsored by Monsanto*" would state Pedro Peretti, FAA Institutional Relations Director, at the AAPRESID Congress, August 22, 2009.

8. The intuition of founders Gustavo Grobocopatel and Víctor Trucco was that technological innovation is endogenous to the agribusiness model, and so profit can be captured from the added value derived from the process of continuous innovation. Harnessed by public policy, it would produce systemic sectoral competitiveness.

9. "Actores ocultos", *Página 12*, May 04, 2008.

10. Presidential discourse (March 29, 2016) available at http://www.casaro sada.gob.ar/informacion/discursos/35865-palabras-del-presidente-mauri cio-macri-en-el-acto-de-inauguracion-del-puente-parada-robles-exaltacion-de-la-cruz

CONCLUSION

Multinational chemical and biotech companies unleashed a revolution in the 1990s that changed the structure of production through the adoption of agrochemicals and GM seeds. The "soybean package" (GM seeds + agrochemicals + direct sowing) has become the paradigm of global agricultural production. Its overriding economic efficiency has given the MNCs superseding power to control other links in the chain. As such, control over technology has become a microeconomic tool of dominance over the chain. More efficient and financed by the companies, producers who adopt the package face cost advantages. This cost-saving opportunity has led to the progressive adoption of this standard in a global scale. By leveraging cost and scale advantages, trading and processing integrated vertically with input providers. This consolidation of links in the soybean chain implied a consolidation of power and control within it. On the supply side, chemical/seeding companies control the inputs by having a de facto monopoly on biotech GM technology. On the demand side, traders/processing companies leverage their scale advantages and exert pressure over producers to establish themselves as dominant buyers.

However, the international mode of agricultural production—specifically, the soybean agribusiness model—was not implanted in a vacuum but in segmented economic spaces, both in social and geographical terms. The transformation had an impact on the domestic agricultural political economy of Brazil, Argentina, and Paraguay, generating local—albeit regionally integrated—identifiable production systems defined as *Soybean Republics*.

© The Author(s) 2017
M. Turzi, *The Political Economy of Agricultural Booms*,
DOI 10.1007/978-3-319-45946-2

Differential institutional structures and different political economy coalitions and conditions processed these external conditions in different ways. Brazil shows an effective pattern of *coordination* between the resource and the institutional structure because the revenue-power equation is more diversified or decentralized. As a result of this context, governors and municipalities are at the forefront of promotion and regulation of the soybean agribusiness mode of production, by establishing working relations with or becoming themselves dominant players in the soybean chain governance. Conflict is focused along the conservation of the Amazon and land tenure, with a clear dominance of landed elites allied with local instances of power, such as governors. Brazilian federalism provided the political flexibility to articulate interest representation locally and at the federal level through the congressional *bancada ruralista*. The Paraguayan political system historically evolved subordinate to specific social groups and private foreign interests. Without institutionalization, conflict time and again degenerates into confrontation, and the one basis for resolving conflict is power, that is, the imposition of the will of those with the organizational, material, and political resources over those without them. The state structure at the local level perpetuates the asymmetric realities between the (foreign, modern agribusiness giant) haves and the (indigenous and landless peasant) have-nots. At best, the result is lack of change and institutional incapacity, and in its worst manifestation, lawlessness due to institutional complicity with the dominant sectors. The expansion of the soybean chain has meant growing territorial control by Brazilian sponsored interests. Producers in Paraguay are mainly Brazilian emigrates (*brasiguayos*), and the benefits of the model do not accrue to the country in terms of inclusive growth or development. In economic terms, soybean production has resulted in a model of low-added value, dependent and enclave-like. Politically, these interests have *colonized* institutions governing the agricultural sector. Argentina is a paradoxical case. Its productive structure does not carry such inherent conflicts as the Brazilian (land concentration, Amazon) or the Paraguayan (Brazilian producers vs. landless and displaced peasants, both carrying harsh social grievances). It also enjoys much more flexible land tenure and rental arrangements, which have led to a more atomized rural structure, which in turn has contributed to the sector's competitiveness and at the same time, a less concentrated mode of political representation. However, political survival of the hyper presidential system in Argentina demands constant distribution of benefits from the part of the executive to its urban/industrial/labor support base.

Economically efficient and politically weak and disorganized, the agricultural sector has—in the context of a commodity boom—constituted a perfect target for the state structure to extract rents and redistribute them to the more concentrated and mobilized base. The model of sector/government relation has been *confrontational*. The agricultural sector is structurally at odds with the political system—in the need for rent capture of the latter—for it produces the rents needed by the Executive to arbiter the distributional game. During the last decade, this tension exploded, aggravated by political mismanagement. Despite recent governmental changes, the built-in institutional imperative to exact rents still remains, and the agricultural sector remains a prime target.

Is the soybean boom would be a re-enactment of the traditional Latin American model of international economic integration as commodity exporter? Should this explanation be the case, the perils that have historically come with this model—international price volatility leading to domestic boom and bust cycles, excessive commodity dependence, and economic enclave features—would still be present dangers. In international political economy terms, soybean production in the BAP countries reflects the emergence of a new global food regime. The rise of living standards throughout the emerging world is generating the traditional gravitational pull for Latin America's commodities. If the region does not find a way out of this model of insertion based on riding successive waves of commodity export booms, it will be left to the consequent busts that lurk at the cusp of those cycles. The evidence for soybeans in South America could be applicable for comparative political economy studies in other countries and products, like US corn, Ukrainian wheat, or Indian cotton.

Further insights are also needed about the costs and benefits of agricultural production. Many of the forward and backward linkages involved in soybean production are still unaccounted for in national sectoral policy analyses, stretching all the way to machinery, chemical, and biotech industries. Downstream, this is also the case with crushing, trading, infrastructure development, and financial services. On the costs side, the effects of soybean production should include its effects on human populations (in terms of health and social conflict), soil degradation, crop rotation, and the environment. In light of these factors, governments in the BAP countries should strive to coordinate policies in order to fit the regionalized economic drivers of this mode of agricultural production. A proactive stance from the governments should be able to capture benefits in this economic segment and channel them toward poverty alleviation and inclusion without impinging on the economic incentives that have driven efficiency and competitiveness. An example of such win–win policies is climbing up the value chain, like the fostering of the oil industry in Argentina. This policy has generated more employment and stimulated linkages in the process of adding value to the raw material. At the same time, the government can capture part of the benefits, improving its fiscal and external position. Harnessing development would not only improve living conditions of the populations, but also avoid generating the initial conditions for populist leaderships to denounce the whole system as inherently exploitative and illegitimate.

ANNEX 1. AGRICULTURAL FORWARD AND BACKWARD LINKAGES

Forward Linkages

A) *Inflation:* Although excessive inflation destroys purchasing power across the board, for the agricultural sector in particular the effects will vary depending on the source of the rise in the general price level. If it is due to an increase in aggregate demand from the rest of the world (*demand-pull*), then it may even be beneficial, as the sector is particularly well positioned to capture extraordinary rents due to exchange rate differentials.[1] However, if the source is a decrease in aggregate supply—an increase of costs, like a rise in wage rates—it will very likely squeeze agriculture returns. Competitive export sectors find it hard to pass on cost increases associated with *cost-push* inflation, and the rents on fixed factors take the adjustment. Thus, inflation takes a particularly hefty toll on land asset values, because expected inflation will tend to move with or ahead of actual inflation. More importantly, it has a distortionary effect on production. Inflationary contexts alter the return calculus of land use: the use value of production decreases and makes more attractive its use as a real estate asset.

B) The *exchange rate* is largely recognized as one of the most important determinant of real agricultural prices, through its determinant effect on price formation. Accepting the real exchange rate as the ratio of the price of tradables to home goods or non-tradables (Dornbusch 1974, 1976), then it is easily observable how an exchange rate appreciation will adversely affect the relative price of agricultural tradables to non-agricultural goods. This increase in

© The Author(s) 2017
M. Turzi, *The Political Economy of Agricultural Booms*,
DOI 10.1007/978-3-319-45946-2

the price of non-agricultural goods (commonly resulting from industrial protection using trade restrictions) will have the "double whammy" effect of lower relative prices of agricultural (export) products and higher cost of agricultural (imported) inputs.

C) *Interest rates:* Their importance for investment decisions in agriculture was already demonstrated by Mundlak (1997). High domestic interest rates raise the cost of capital for a sector highly dependent on debt finance and put upward pressure on the costs of agricultural production. Agriculture is more sensitive than non-agriculture to changes in interest rates and less sensitive to changes in the cost of labor. According to Mundlak, capital/labor ratio in the sector increases over the long run, so rises in interest rates are particularly damaging to agricultural investment, a proxy of future returns. Moreover, high interest rates reduce land asset values by raising the discount rate on future income flows and also reduce the incentives for holding stocks of grain crops, dislocating present and future supply flows and putting downward pressure on prices. If higher nominal interest rates signal higher inflation expectations, then an outflow of capital from the sector to more secure assets is likely to take place.

D) *Government taxing and spending policies* (fiscal policy) reveal a more direct forward linkage. Broadly, a higher tax burden on labor incomes and on expenditures would make capital a relatively cheaper factor of production, thus favoring the relatively capital-intensive agricultural sector. Exemptions from capital gains tax and investment allowances are especially favorable to agricultural investment. Measures like export grants, reduced loan rates, and storage support actually transfer wealth from taxpayers to agriculture. Historically, the governmental tilt toward industry meant the protection of domestic manufacturing. This restriction of trade implied a panoply of measures that included subsidies for input, soft credits, tariffs, price controls, quantitative restrictions, exchange, and capital controls. These policies had unintended yet strong effects on agriculture: they modified investment incentives for agriculture vis-à-vis other sectors through its effects on the exchange rate; triggering the "double whammy" effect mentioned earlier.

E) *International markets:* Their conditions also significantly affect performance and drive behavioral responses in the national agricultural sectors. Shifts in prices in the international commodity markets tend to be stark because prices adjust to equilibrate relatively inelastic supply and demand. Moreover, world spot markets—as export destinations or as import substitutes—often set agricultural commodity price levels for domestic producers, since the "law of one price" holds for fairly homogeneous bulk agricultural commodities (Richardson 1978). This is why producers typically push for measures like price leveling and averaging, because they cushion the negative effects of spot market price formation and its consequent volatility of agricultural prices at the retail price level. Where international trade is important—as is the case for the countries under study—overseas prices are found to be significant determinants[2] of domestic commodity prices (Bale and Lutz, 1979). For example, when agricultural profits are high, the sector attracts investment. The result—*ceteris paribus*—is increased production. But as production rises, prices fall, because bulk commodity demand does not increase as prices drop. Agricultural world markets are further distorted by subsidies from developed countries, supporting costs of production even when prices fall below production cost. As a result, lower prices do not lead to decreased output and hence prices tend to fluctuate considerably. Producers can then react to falling prices by reducing costs, spreading across a greater area or increasing volumes. This has the agricultural sector in an "expand or die" cycle, while at the same time it has produced constant productivity gains derived from advances in equipment, seeds, chemicals, cultivation techniques, and management advisory services.

Backward Linkages

A) Agricultural products are subject to an unpredictable and unpreventable degree of variability due to climatic influences. These seasonal fluctuations translate directly into international prices. Agricultural prices are quicker to adjust than industrial prices due to the almost perfect competition that arises from broad product homogeneity.[3] In order to mitigate these negative effects, the agricultural sector has spurred the generation of a variety of *financial instruments* for hedging and insurance purposes, such as futures markets.

B) Where the agricultural sector is export oriented, the contribution to the *trade balance* tends to be significant in terms of share of total exports and import substitutes. The exports of primary products as percentage of total exports are 69% for Argentina, 51.4% for Brazil, and 84% for Paraguay.[4] Agricultural products have represented on average 7% of GDP in Argentina, 5% in Brazil, and 19% in Paraguay for the period 1990–2008. The monetary linkages manifest in the capital component of the balance of payments, where the export competitive agricultural sector has historically contributed a sizeable portion of foreign exchange reserves. This has relieved pressure to finance the industrial sector—typically more exposed to import competitors—and cushioned state budgets. Governments have found ways—from taxes and export duties to the nationalization of grain trade—to appropriate some part of the extraordinary sector returns to improve fiscal standing and strengthen political coalitions.

C) In the past, agriculture was a *labor*-intensive activity, much more so than cattle breeding. This was coupled with geographic dispersion of production, which required linkages with the communication and transportation sectors, demanding *infrastructure* development. For the last 30–40 years, *technology* has also played a determinant role in the transformation of the agricultural chain of production. Extensive R&D investment has led to mechanization upgrades, advances in biotech and seeding techniques, transforming traditional agriculture into non-labor-intensive agribusiness. Large-scale technology biases have had on the one side labor-saving effect, but on the other they have skyrocketed agricultural productivity. Although it remains a contentious issue, it seems fair to say that agriculture has contributed to an overall rise of national income and a release of labor resources for use in the rest of the economy, an especially critical point in the initial take-off phases[5] of modern growth.

D) A distinctive feature of the agricultural sector is that production rests on a fixed asset with limited availability: *land*. Despite being somewhat toned down by productivity increase possibilities due to innovation in seeding and cultivation techniques, land access and use still remain the single most important link in the agricultural chain, with reverberations throughout the entire economic structure. Land is both a production input and an asset. The institutional structure governing land ownership and rental has a significant influence on what use it will be giving, thereby shaping production decisions and output. Incentives for investments in land improvements or the prospect of financing (using land as collateral) are largely shaped by property rights. Argentina, Brazil, and

Paraguay present three very different land regimes, which are explored in-depth in the case studies.

E) Finally, agriculture—like no other economic sector—has *environmental* linkages. Natural resources' net worth exceeds their direct production value, calculable in terms of biodiversity or carbon retention. In the absence of markets for these services, these additional values/costs of natural resources are not internalized by farmers. Profit calculation is then based solely on the market value, neglecting long run economic and—even if beyond the scope of this study—ecological sustainability.

NOTES

1. This effect was somehow canceled in the BAP countries (especially Argentina) because the government imposed a different exchange rate for exporters in order to discourage exports.
2. Other exogenous factors that impact the agricultural sector are international capital transfers and world real interest rates.
3. Considering that commodity cartels, trading stock exchanges, storage conditions, and government policies influence price determination, it seems a reasonable assumption that to a large extent agricultural markets do not operate under product differentiation, and hence in conditions akin to perfect competition.
4. Economic Commission for Latin America and the Caribbean (ECLAC), *Statistical yearbook for Latin America and the Caribbean 2008.*
5. Arthur W. Lewis proposed in his *Theory of economic growth* (1955) the concept of "dual economy" or "two sector model". The model describes an economy with a traditional sector—with low wages and a nearly infinite supply of labor—and the modern, capital abundant sector. Assuming a zero marginal labor productivity for the traditional sector, the process of structural change is powered by the rural and urban marginal utility differential that causes rural laborers to leave agriculture due to the relatively higher industrial wage rates.

ANNEX 2. GEOGRAPHIC DISTRIBUTION OF SOYBEAN PRODUCTION IN THE SOYBEAN REPUBLICS

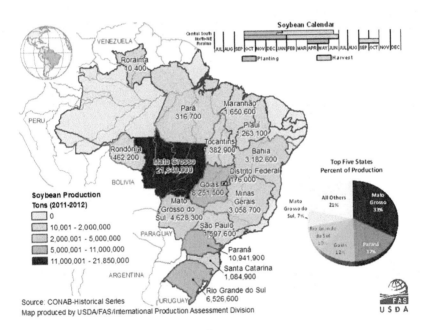

Map A.1   Brazil: soybean production by state

© The Author(s) 2017
M. Turzi, *The Political Economy of Agricultural Booms*,
DOI 10.1007/978-3-319-45946-2

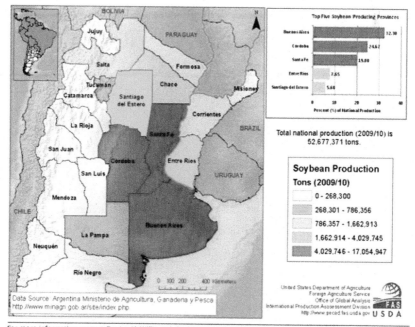

Map A.2   Argentina: soybean production by province

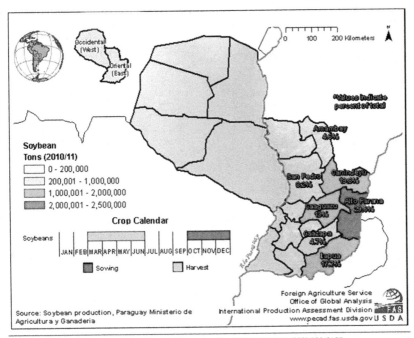

Map A.3 Paraguay: soybean production by province

References

Alegre Sasiain, Pedro, Efraín and Aníbal, Orué Pozzo. 2008. *La tierra en Paraguay 1947-2007: 60 años de entrega del patrimionio nacional. Stroessner y el Partido Colorado.* Asunción: Arandurã Editorial.

Alston, Lee, Gary Libecap and Bernardo Mueller. 1999. *Titles, conflict and land use: the development of property rights and land reform on the Brazilian Amazon Frontier.* Michigan: University of Michigan Press.

Balassa, Bela. 1971. *The structure of protection in developing countries.* Baltimore: World Bank and Inter-American Development Bank, Johns Hopkins Press.

Balisacan, Arsenio and J.A. Roumasset. 1987. Public choice of economic policy: The growth of agricultural protection. *Weltwirtschaftliches Archiv* 123: 232–248.

Bale, Malcolm and E. Lutz. 1979. The effects of trade intervention on international price instability. *American Journal of Agricultural Economics* 61: 512–516.

Baletti, B. 2014. Saving the Amazon? Sustainable soy and the new extractivism, *Environment and Planning A*, Vol 4.

Barros, Gerlado 2009. Brazil: The challenges in becoming an agricultural super-power. In *Brazil as an Economic Superpower? Understanding Brazil's Changing Role in the Global Economy*, ed. Lael Brainard and Leonardo Martinez-Diaz, Washington, DC: Brookings Institution Press.

Barsky, Osvaldo and Mabel Dávila. 2008. *La rebelión del campo. Historia del conflicto agrario argentino.* Buenos Aires: Sudamericana.

Bhagwati, Jagdish. 1978. *Foreign trade regimes and economic development: Anatomy and consequences of exchange control regimes.* Cambridge: Ballinger Press.

Becker, Gary S. 1983. A theory of competition among pressure groups for political influence. *Quarterly Journal of Economics*, 98 (3).

© The Author(s) 2017   141
M. Turzi, *The Political Economy of Agricultural Booms*,
DOI 10.1007/978-3-319-45946-2

Bhardwaj, Geetesh and Gorton, Gary B. and Rouwenhorst, K. Geert. 2015. Facts and Fantasies About Commodity Futures Ten Years Later Yale ICF Working Paper No. 15–18.

Bisang, Roberto and Graciela E. Gutman 2005. The accumulation process and agrofood networks in Latin America. *Cepal Review* 87: December.

Bueno de Mesquita, Bruce. 2003. *The logic of political survival.* Cambridge: MIT Press.

Bulmer-Thomas, Victor. 2003. *The economic history of Latin America since independence.* Cambridge: Cambridge University Press.

Buchanan, James M. and Gordon Tullock. 1962. The Calculus of Consent: Logical Foundations of Constitutional Democracy, Ann Arbor Paperbacks.

Bullock, David and K.S. Jeong. 1994. A critical assessment of the political preference function approach in agricultural economics: comment. *Agricultural Economics* 10: 201–206.

Cadot, Olivier, Antoni Estevadeordal, Akiko Suwa-Eisenmann, and Thierry Verdier. 2004. *The Origin of Goods: A Conceptual and Empirical Assessment of Rules of Origin in PTAs.* Washington: IADB- and CEPR.

Carter, Miguel. 2015. Epilogue. Broken promise: The land reform debacle under the PT governments. In *Challenging social inequality: The landless rural workers movement and Agrarian reform in Brazil*, ed Miguel Carter. Duke University Press.

Carter, Miguel and Eduardo Zegarra. 2000. Land markets and the persistence of rural poverty in Latin America: post-liberalization policy options. In *Rural poverty in Latin America: Analytics, new empirical evidence and policy*, ed. Alberto Valdes and Ramon Lopez. Basingstoke: Macmillan.

Clapp, J. 2015. Distant agricultural landscapes, *Sustainability Science*, April 2015, Volume 10, Issue 2.

Corrales, Javier. 2002. *Presidents without parties: The politics of economic reform in Argentina and Venezuela.* Pennsylvania: Pennsylvania State University Press.

Costa, Sandra. 2015. The Agrarian question in Brazil and the caucus in Congress. Unpublished thesis, FFLCH-USP.

Christensen, Clayton. 1997. *The innovator's dilemma: When new technologies cause great firms to fail.* Harvard Business School Press.

De Gorter, Harry and Swinnen, Johan. 2002. Political economy of agricultural policy. In *Handbook of Agricultural Economics*, ed. Bruce Gardner and Gordon Rausser. Vol. 2 chapter 36.

Dornbusch, Rudiger. 1974. Tariffs and non-traded goods. *Journal of International Economics* May: 177–185.

Dornbusch, Rudiger. 1976. Expectations and exchange rate dynamics. *Journal of Political Economy* 84: 1161–1176.

Downs, Anthony. 1957. *An economic theory of democracy.* New York: Harper and Row.

Drucker, Peter. 1992. *The age of discontinuity: Guidelines to our changing society.* Boston: Harper & Row.

Etchemendy, Sebastián, and Ruth Collier. 2007. Down but not out: Union resurgence and segmented neocorporatism in Argentina. *Politics & Society* 35(3): 363–401.

Francois, J. and Woerz. 2008. Producer Services, Manufacturing Linkages, and Trade, *J. J Ind Compet Trade*, 8: 199.

Fuck, Marcos Paulo. et al. 2008. Intellectual property protection, plant breeding and seed markets: a comparative analysis of Brazil and Argentina. *International Journal of Technology Management and Sustainable Development* 7: 3.

Fogel, Ramón. 1989. *La cuestion agraria en el Paraguay: apuntes para su estudio.* Itapúa: Estudios Rurales de Itapúa.

Fogel, Ramón 2008. Agronegocios, conflictos agrarios y soberanía alimentaria en el Paraguay. *8th Meeting of the Rural Development Working Group, Latin American Council of Social Sciences* (CLACSO).

Fogel, Ramón and Marcial Riquelme eds. 2005. *Enclave sojero: merma de soberanía y pobreza.* Asunción: Centro de Estudios Rurales. Interdisciplinarios.

Foucault, Michel. 1977. *The archaeology of knowledge and the discourse on language.* California: Vintage Press.

Galeano, Eduardo. 1970. *Las venas abiertas de América Latina.* Montevideo: Catálogos.

Galeano, Luis. 2012. Paraguay and the expansion of Brazilian and Argentinian agribusiness frontiers. *Canadian journal of development studies/Revue canadienne d'études du développement* 33(4): 458–70.

Gardner, Bruce and Gale Johnson. 2002. Policy-related developments in agricultural economics: Synthesis of handbook volume 2. In *Handbook of agricultural economics*, ed. Bruce Gardner and Gordon Rausser. Vol. 2, Chapter 42, 2215–2249.

Garrett Rachael D. and Lisa L. Rausch. 2016. Green for gold: social and ecological tradeoffs influencing the sustainability of the Brazilian soy industry. *The Journal of Peasant Studies*, Vol. 43, No. 2.

Gereffi, Gary. 1999. *A commodity chains framework for analysing global industries.* Duke University.

Gereffi, Gary and Miguel Korzeniewicz. 1994. *Commodity chains and global capitalism.* Westport, CT: Praeger.

Germani, Gino. 1965. *Política y sociedad en una época de transición: de la sociedad tradicional a la sociedad de masas.* Buenos Aires: Paidós.

Hartle, Douglas. 1983. The theory of 'rent seeking': some reflections. *Canadian Journal of Economics* 16: 539–554.

Hirschman, Albert O. 1958. *The strategy of economic development.* New Heaven: Yale University Press.

Hopkins, T. K. and Wallerstein, I. (1994). Commodity chains: construct and research, in Gereffi, G. and Korzeniewicz, *Commodity Chains and Global Capitalism.* Westport: Greenwood Press.

Horst, René Harder. 2007. *The Stroessner regime and indigenous resistance in Paraguay.* Gainesville: Florida University Press.

Hymowitz, Theodore. 1970. On the domestication of the soybean. *Economic Botany* 24(4): 408–421.

Isern Munne, Pedro. 2007. *The distribution struggle in Argentina: Introduction to the agricultural sector's political economy.* Fundación Pensar, Buenos Aires.

Iversen, Torben. 2006. Democracy and Capitalism. In: Wittman D, Weingast B *Oxford Handbook of Political Economy.* Oxford; New York: Oxford University Press; 2006. pp. 601–623.

Krueger, Anne. 1978. *Foreign trade regimes and economic development: Liberalization attempts and consequences.* Cambridge, MA: Ballinger.

Krueger, Anne 1993. Virtuous and vicious circles in economic development. *American Economic Review* May.

Krueger, Anne. 1996. Political economy of agricultural policy. *Public Choice* 87: 173–175.

Krugman, Paul R. and Maurice Obstfeld. 2003. *International Economics: Theory and Policy.* Addison Wesley, NY.

Laino, Domigo. 1977. *Paraguay: fronteras y penetración Brasilera.* Asunción: Ediciones Cerro Corá.

Lapola, D.M., L.A. Martinelli, C.A. Peres, J.P.H.B. Ometto, M.E. Ferreira, C.A. Nobre, A.P.D. Aguiar et al. 2013. Pervasive transition of the Brazilian land-use system. *Nature Climate Change* 4(1): 27–35.

Lebow, Richard Ned. 2007. *Coercion, cooperation, and ethics in international relations.* London: Routledge.

Levitsky, Steven and Maria Victoria Murillo. eds. 2005. *Argentine democracy: The politics of institutional weakness.* University Park: Pennsylvania State University Press.

Macours, Karen, Alain de Janvry and Elizabeth Sadoulet. 2004. Matching in the Tenancy Market and Access to Land. *CUDARE Working Papers*, Department of Agricultural & Resource Economics, University of California, Berkeley.

Martinelli, L.A., R. Naylor, P.M. Vitousek, and P. Moutinho. 2010. Agriculture in Brazil: Impacts, costs, and opportunities for a sustainable future. *Current Opinion in Environmental Sustainability* 2(5–6): 431–438.

McDonald, James H. 1999. The neoliberal project and governmentality in rural Mexico: emergent farmer organization in the Michoacán Highlands. *Human Organization* 58(3): 274–284.

Muller, Gerlado. 1989. *Complexo Agroindustrial e modernização agrarian.* São Paulo: Hucitec, Educ.

Mundlak, Yair. 1997 The dynamics of agriculture. In *Proceedings of the XIII international conference of agricultural economics*, Sacramento, CA, August 10–16.

Murillo, María Victoria and Andrew Schrank. 2009. Labor unions in the policy-making process in Latin America. In *Actors in Latin American Policymaking*, ed. Ernesto Stein and Mariano Tommasi. Inter-American Development Bank.

Murillo, María Victoria. 2008. From Kirchner to Kirchner. *Journal of Democracy*, 19(2).

Murphy, S. 1999. Market power in agricultural markets: some issues for developing countries. *South Centre Working Paper*, Geneva.

Murphy, S., et al. 2012. Cereal Secrets: The Worlds Largest Grain Traders and Global Agriculture. *OXFAM Research Reports*, Oxford, UK.

Neupert, Ricardo. 1991. Brazilian colonization in the Paraguayan agricultural frontier. *Notas de Población* 18–19(51–52): 121–154.

Nickson, Andrew 1981. Brazilian colonization of the Eastern border region of Paraguay. *Journal of Latin American Studies* 13: 111–131.

North, Douglass. 1990. *Institutions, institutional change, and economic performance*. Cambridge: Cambridge University Press.

Nurkse, Ragnar. 1962. *Patterns of trade and development*. London: Blackwell.

O'Donnell, Guillermo. 1978. State and alliances in Argentina, 1956–1976. *Journal of Development Studies* 15(1): 3–33.

Oliveira, Gustavo de L. T. 2016. The geopolitics of Brazilian soybeans. *The Journal of Peasant Studies* 43(2): 348–372.

Olson, Mancur. 1990. Agricultural exploitation and subsidization: There is an explanation. *Choices* 4: 8–11.

Ordeshook, Peter C. 1990. The emerging discipline of political economy. In *Perspectives on positive political economy*, ed. James Alt and Kenneth Shepsle, 9–30. Cambridge: Cambridge University Press.

Peláez, Victor and Poncet Christian. 1999. Estratégias industriais e mudança técnica: uma análise do processo de diversificação da Monsanto. *História Económica & História de Empresas* II(2).

Peltzman, Sam. 1976. Toward a More General Theory of Regulation, *Journal of Law and Economics*, Vol. 19, no. 2.

Raikes, Philip, Michael Jensen and Stefano Ponte. 2000. Global commodity chain analysis and the French filière approach: comparison and critique. *Economy and Society* 29(3): 390–417.

Reydon, B. 2000. Intervenção estatal no mercado de terras: a experiência recente no Brasil. In *Estudos NEAD*, Vol. 3. Ministério do Desenvolvimento Agrário: Rio de Janeiro.

Richardson, J. David. 1978. Some empirical evidence on commodity arbitrage and the law of one price. *Journal of International Economics* 8: 341–351.

Riker, W. 1980. Implications from the Disequilibrium of Majority Rule for the Study of Institutions. *The American Political Science Review*, 74(2).

Robinson, James A. 1999. When is a state predatory?, *CESifo Working Paper Series No. 178.*

Robinson, James A, Ragnar Torvik and Thierry Verdier. 2006. Political foundations of the resource curse. *Journal of Development Economics* 79: 447– 468.

Rock, David. 2002. *State building and political movements in Argentina, 1860–1916.* California: Stanford University Press.

Roett, Riordan. 1989. Paraguay after Stroessner. *Foreign Affairs*, Spring.

Roett, Riordan and Richard Sacks. 1991. *Paraguay: The personalist legacy*, Boulder, San Francisco, Oxford: Westview Press.

Scherer, F.M. and Ross, David. 1990. Industrial Market Structure and Economic Performance, University of Illinois at Urbana-Champaign's Academy for Entrepreneurial Leadership Historical Research Reference in Entrepreneurship, NC.

Schneider, Ben Ross. 2004. *Business politics and the state in twentieth-century Latin America.* Cambridge: Cambridge University Press.

Shurtleff, William and Akiko Aoyagi. 2007. History of Soybean crushing: Soy oil and soybean meal. In *History of Soybeans and Soyfoods: 1100 B.C. to the 1980s,* Lafayette, CA: Soyinfo Center.

Stigler, G. 1971. The Theory of Economic Regulation. *The Bell Journal of Economics and Management Science*, 2(1), 3–21.

Stein, Ernesto and Mariano Tommasi. 2007. The institutional determinants of state capabilities in Latin America. In *Annual World Bank Conference on Development Economics Regional: Beyond Transition*, ed. Francois Bourguignon and Boris Pleskovic. Washington, DC: World Bank.

Steward, C. 2007. From colonization to "environmental soy": A case study of environmental and socio-economic valuation in the Amazon soy frontier, *Agriculture and Human Values*, Vol 24, Issue 1.

Swinnen, J. F.M. 2009. Reforms, globalization, and endogenous agricultural structures. *Agricultural Economics*, 40: 719–732.

Tang, Ke and Wei Xiong. 2010. Index Investment and Financialization of Commodities, NBER Working Paper No. 16385.

Teece, David J. 2000. *Managing intellectual capital.* Nova York: Oxford University Press.

Toledo, Marcio Roberto. 2005. *Circuitos espaciais da soja, da laranja e do cacau no Brasil: uma nota sobre o papel da Cargill no uso corporativo do território brasileiro*, Ph.D. book, Instituto de Geociências, Programa de Pós Graduação em Geografia, Universidade Estadual de Campinas, Brazil.

Topik, Steven, Carlos Marichal and Zephyr Frank. 2006. *From silver to cocaine: Latin American commodity chains and the building of the world economy, 1500–2000.* Durham, NC: Duke University Press.

Vigna, Edélcio. 2007. Bancada Ruralista: o maior grupo de interesse no Congresso Nacional. In *Argumento*, Vol. 12. Brasília: INESC.

Wilkinson, John 1997. Regional integration and the family farming the Mercosul countries. In *Globalizing food: agrarian questions and global restructuring*, ed. David Goodman and Michael Watts. London: Routledge.

Williamson, John. 2005. *Curbing the boom-bust cycle: Stabilizing capital flows to emerging markets*. Washington, DC: Institute for International Economics.

Williamson, Jeffrey G, Kevin H. O'Rourke and David S. Jacks. 2009. Commodity price volatility and world market integration since 1700. NBER Working Paper 14748, February.

Wesz Jr, Valdemar. 2016. Strategies and hybrid dynamics of transnational soy companies in South America, *The Journal of Peasant Studies*, 43:2.

INDEX

A

ABC Color, 111
ABRASEM, 42
Aceitera General Deheza (AGD), 57, 64
Agrarian Federation (FAA), 123
Agribusiness Chambers Association (ACTA), 123
Agricultural and Forestry Biosecurity Commission (CBAF), 36
Agricultural Cooperative Confederation (CONINAGRO), 123
Agricultural Coordination of Paraguay (CAP), 39, 107
Agriculture
    global trade, 50–55
    infrastructure, 120, 129, 134
    Latin American economies, 8–10
    macroeconomics of, 14–18
Alfonsín, Raúl, 12
AMaggi, 63, 65, 77
Amazon, 85–89, 128
APROSEMP, 39
APROSOJA, 43–44, 86
Aquino, Antonio Ibáñez, 36
Archer Daniels Midland (ADM), 35, 54, 57, 59, 61, 77

Argentina
    agricultural reforms, 51
    agrochemical markets, 27
    biosafety laws, 33
    confrontation, 117–125
    corporate production strategies, 66
    CTNBio, 41
    ETs, 72
    FDI inflows, 53
    federal policy, 87
    of financial intermediation, 67–68
    geopolitics, 103, 105–106, 112
    GM seeds, 25, 35
    land tenure, 93
    livestock farming, 64
    patent laws, 34
    during Perón, 84
    political economy, 118–119, 127–128
    R&D spending, 38
    RR soybeans, 36
    rural sector, 91–92
    seed industry, 27, 31–32
    soybean plantations, 5, 11–14
    trading companies, 57, 60–62
    trading, soybean, 58–59
    vegetable oil industry, 63, 130

© The Author(s) 2017
M. Turzi, *The Political Economy of Agricultural Booms*,
DOI 10.1007/978-3-319-45946-2

Argentina–Brazil Integration and
    Economics Cooperation Program
    (PICE), 12
Argentine Association of Regional
    Consortiums for Agricultural
    Experimentation (AACREA), 123
Argentine No-till Farmers Association
    (AAPRESID), 123
Argentine Rural Confederation
    (CRA), 123–124
Association of Soy Growers of
    Paraguay (APS), 38–39

B

Bancada ruralista (BR) or
    congressional rural bloc, 89–91,
    97, 122, 128
BAP countries, 8, 13, 18, 33, 49–51,
    55, 66, 72, 129
BASF, 26–27, 29
Bayer, 26–27
Biosafety Bill (N. 11.105), 40
Biosecurity Commission
    (COMBIO), 36
Biotech companies, institutional
    framework
  economic, 29–30
  legal, 28–29
  political, 30
  scientific, 30
Biotechnology, 25, 36–37
Bolsa de Cereais e Mercadorias do
    Estado de Mato Grosso, 67
Brasiguayos, 103–106, 111, 128
Brasiguayo, see Brasiguayos
Brazil
  agribusiness, 18
  coordination patterns, 83–97,
    127–128
  FDI inflows, 53
  financial infrastructure, 66–68
  geoeconomic dynamics, 74–77
  geopolitics, 102–103, 105–106,
    108, 112
  GMOs regulations, 40, 42–44
  GM seeds, 25, 33, 36–37
  grain program, 39
  land tenure, 94–97
  political economy, 85–89, 91, 93
  R&D spending, 27, 38
  rural sector, 117
  seed industry, 32
  soybean plantations, 5, 11–14
  taxes and duties, 71, 122
  trading companies, 57–65
Brazil, Argentina, and Paraguay (the
    BAP countries), 8–9, 13, 18, 28,
    33, 49–51, 54–56, 61–62, 66–67,
    69–70, 72, 123, 129
Brazilian consumer rights association
    (IDEC), 40–41
Brazilian Institute of Geography and
    Statistics (IBGE), 92–93
Buenos Aires Grain Exchange in
    Argentina (MAtBA), 67–68
Bunge, 35, 54, 57, 61

C

Caramuru Alimentos, 64
CARBAP, 123
Cargill, 29, 32, 35, 54, 57–61, 76–77
Cartes, Horacio, 71, 75, 108, 112
Chicago Board of Trade (CBOT), 59,
    67–68
Clientelism, 102
Colonization, 38, 39, 95, 101,
    104–108, 112
Commodity chain, 14–18, 28
Commodity chain analysis (CCA), 15
Confederation of Agriculture and
    Livestock (CNA), 89–90
Confrontation, 31, 43, 118, 128–129

D
Decree 7797/00, 37
DEM (Democrats), 90
Don Mario, 34
Dow Chemical Co., 26–27
Duhalde, 73, 120
DuPont Co., 26–27, 29, 42

E
El Tejar, 64
Embrapa, 31–32, 35, 42–43, 92
ETC Group, 26, 29

F
Farmer
  Brazilian, 85–86, 88,
    94, 106
  collective actions, 41, 44
  GM seed, 34–35, 37–39
  infrastructure, 67, 69
  land tenure, 94
  market power, biotech
    companies, 28–30
  Paraguayan, 107
  political economy models, 17–18
  protests and strikes, 73
  relocations, 103
  scientific feed formulation, 6
  state intervention, 52–57
  subsidies, 96
  technical developments, 24–25
Farmer's Union Syndicate
  (UGP), 38–39, 107, 110–111
Federal Solidarity Fund
  (FFS), 73
Federation of Agricultural Workers
  (FETAG), 44
Federation of Agriculture and
  Livestock of the State of Mato
  Grosso (FAMATO), 44, 86

Federation of Cooperatives of
  Production (FECOPROD), 39
Feed, 1–3, 6–8, 58, 66, 76
Fight for Land Organization
  (OLT), 107, 110
Finance, 1, 4, 8, 10, 15, 25, 50, 54,
  57–58, 62, 66–67, 102, 127,
  132, 134
Food products, 1–9, 25, 29, 33,
  40, 50, 55–57, 83–85, 95,
  124, 129
Forward contracts (FCs), 68–69
Fuel, 1, 3, 24, 41, 75, 96

G
Genetically modified organism
  (GMO), 28–30, 33, 36, 38–40
Genetic use restriction technology
  (GURT), 30
Glyphosate, 24, 35, 43
GM seed, 34–35, 37–39
Grain swaps, 69
Greenpeace, 40–41
Gross domestic product (GDP), 18,
  38, 70, 112, 134
Growth Acceleration Program
  (PAC), 41

H
Harvest, 13, 24, 40, 43,
  56–57, 69

I
ICMS (Imposto sobre Circulacao de
  Mercadorias e Servicos), 70–71
IDB, 62, 106
INBIO (Instituto de Biotecnología
  Agricola), 38

Infrastructure
  agricultural production, 120, 129,
    134
  Amazonian waterway system, 65, 84
  Brazilian, 76
  feedlot, 6
  geopolitical strategy, 12
  global trading, 50, 55
  logistical, 74–75
  networks, 66–67
  Paraguayan, 106
  public sector investment, 41–42
  storage facilities, 69–70
Institute of Rural Welfare (IBR), 103,
  109
Intellectual property right (IPR), 27,
  29, 35
International Service for the
  Acquisition of Agri-biotech
  Applications (ISAAA), 26, 30

J
Judiciary, 95–97, 109

K
Kirchner, Cristina Fernández de, 73,
  117, 121–122
Knowledge-based economy, 94

L
Land and Rural Development Institute
  (INDERT), 109–110
Landless Movement (MST), 95–96
La Tijereta, 34
Law on Seeds and Protection of Crops
  385/94, 37
Los Grobo, 63–64, 93
Lousteau, 73

M
Macri, Mauricio, 35, 73, 117,
  124–125
Mato Grosso
  AMaggi, 63, 65
  Caramuru Alimentos, 64–65
  deforestation, 85–88
  financial intermediaries, 67
  indigenous populations, 96
  infrastructure, 76–77
  MNCs, 69
  producers, 58–60
  RR technology, 44
  soybean production, 12–13
  taxes and duties, 71
  Xingu Indigenous Park, 89
M&FBOVESPA S.A., 67
Ministry of Agriculture (MAG), 36,
  68, 103
MNCs, 17–18, 31–32, 40, 43, 52,
  68–69, 74, 101, 127
Molinos Rio de la Plata, 64
Monsanto, 24–27, 29, 33–37, 39–40,
  42–44, 56, 114, 123

N
National Advisory Committee on
  Agricultural Biosafety
  (CONABIA), 33
National Agricultural Technology
  Institute (INTA), 31–32
National Biosafety Council
  (CNBS), 40–41
National Control Office for Tobacco
  and Cotton, 36
National Coordinator of Peasant
  Organizations (MCNOC), 107,
  110
National Direction of Plant
  Protection, 36

National Independent Aboriginal
 Organization (ONAI), 107–108
National Institute for Colonization
 and Agrarian Reform
 (INCRA), 95–97
National Institute for Spatial Research
 (INPE), 85–86
National Peasant Federation
 (FNC), 108
National Peasant Union (UCN), 107
National Plant Variety Protection
 Service (SNPC), 42
National Registry of Cultivar Property
 (RNPC), 33
National Registry of Cultivars
 (RNC), 33, 42
National Registry of Plant Varieties
 (RNC), 42
National Registry of Seeds and Plants
 (RENASEM), 42
National Seed Direction, 36
National Seed Institute (INASE), 33
National Service of Plant Health
 Quality (SENAVE)., 36
National Technical Commission of
 Biosafety (CTNBio), 36, 40–41
National Transport Infrastructure
 Department, 86
Nidera, 33–34, 37, 63, 65
Noble Group, 62, 65
Non-GMO seeds, 29
No-till sowing, 24

O
Options on futures, 69

P
Paraguay
 ADM, 59

agricultural governance
 structure., 87
Bunge, 62
Cargill, 59
colonization, 101–112
cotton export, 10
El Tejar, 64
financial intermediation, 67–68
GM seeds, 25, 36–39
infrastructure developments, 74–76
institutional structure, 32
land concentration, 92–94
Louis Dreyfus Group (LDC), 60–61
national champions, 65
Noble Group, 62
political system, 127–128
producer's revenue, 122
regional forces, 90–91
RR technology, 33
sectoral interests, 117–118
soybean production, 12–14
taxes and duties, 70–71
Tierra Roja, 63
waterway system, 66
Paraguayan Chamber of Exporters and
 Traders of Grains and Oilseeds
 (CAPECO), 38–39, 107, 110
Paraguayan Institute of Agrarian
 Technology (IPTA), 38
Paraguayan Peasant Movement
 (MCP), 107
Paraguayan People's Army (EPP), 110
Paraguay stock exchange
 (BVPASA), 68
Pastoral Land Commission (PLC), 87
Peasants, 10, 67, 93–96, 101–103,
 105—112, 117, 128
Peronist party, 118–119
PMDB (Brazilian Democratic
 Movement Party), 90
Political economy
 Amazonian rainforest, 85–97

Political economy (*cont.*)
  commodity chain, 14–18, 28
  financial structure, 66–73
  infrastructure development, 74–77
  international, 101–112
PSDB (Brazilian Social Democracy
  Party), 90–91
PT (Worker's Party), 90
PTB (Brazilian Labor Party), 90

R
R&D initiatives, 25–29, 31–32, 34,
  38, 45, 56, 66, 123
Rosario Futures Exchange
  (ROFEX), 67–68
Roundup, 24–25, 29, 35, 43
RR (Roundup Ready)
  soybean, 24–25, 29, 33–34,
  36–37, 39–40, 43–45
Rural Society (SRA), 123

S
SAGPyA, 33–34
Santa Rosa, 34
Sarney, José, 12
Seed Association (ASA), 123
Seeds and Plants National System Law
  of 2003, 42
Silva, Marina, 41
Soybeans
  ET on, 71, 73, 120, 124
  global trade, 55–65
  technological development, 1, 6–7,
    11, 15, 23–32, 34–41, 43–45, 49,

  52–57, 63, 67–69, 84, 94, 106,
    123, 125, 127, 133–134
  world market, 4–8
  See also BAP countries
Spreads or straddles, 69
SPS, 34
State Development Bank
  (BNF), 106
Stroessner, Alfredo, 102–103,
    105–107, 109, 112
Syngenta, 26–27,
    29, 34, 42

T
Tax system, 70–73
T-GURT "traitor" seeds, 30
Treaty of Itaipú, 105

U
Union of Production Trades
  (UGP), 110–111

V
V-GURT "terminator", 30
Vicentín, 35, 64

W
White bag seeds, 33
Worker Peasant Front (FOC), 108
World Bank, 2, 68, 106
World Census of Agriculture, 93

CPSIA information can be obtained
at www.ICGtesting.com
Printed in the USA
LVOW06*0248061017
551391LV00019B/536/P